DRAIN FOR GAIN: MAKING WATER MANAGEMENT WORTH ITS SALT

Promoter: Prof. dr. ir. Bart Schultz, PhD, MSc
                      Professor of Land and Water Development
                      UNESCO-IHE
                      The Netherlands

Co-promoter: Prof. dr. Wim Cofino
                      Professor of Integrated Water Resource Management
                      Wageningen University
                      The Netherlands

Awarding Committee: Prof.dr.ir. Sjoerd E.A.T.M. van der Zee
                      Wageningen University
                      The Netherlands

                      Prof.dr.ir. Nick C. van de Giesen
                      Delft University of Technology
                      The Netherlands

                      Dr.ir. Pieter J.M. de Laat
                      UNESCO-IHE
                      The Netherlands

                      Prof. Chandra A. Madramootoo Ph.D., Ing.
                      McGill University
                      Canada

                      Dr. Muhammad Nawaz Bhutta
                      International Waterlogging and Salinity Research Institute
                      Pakistan

This research is conducted within the Research School WIMEK-SENSE.

DRAIN FOR GAIN

MAKING WATER MANAGEMENT WORTH ITS SALT

Subsurface Drainage Practices in Irrigated Agriculture
in Semi-Arid and Arid Regions

DISSERTATION
Submitted in fulfillment of the requirements of
the Academic Board of Wageningen University and
the Academic Board of UNESCO-IHE Institute for Water Education
for the Degree of DOCTOR
to be defended in public on Friday, 16 January 2009
at 16:00 hours in Wageningen, The Netherlands

by

Hendrik Pieter Ritzema
born in Baarn, the Netherlands

CRC Press/Balkema is an imprint of the Taylor & Francis Group, an informa business

Published by:
CRC Press/Balkema
PO Box 447, 2300 AK Leiden, The Netherlands
e-mail: Pub.NL@taylorandfrancis.com
www.crcpress.com – www.taylorandfrancis.co.uk – www.balkema.nl

ISBN 978-0-415-49857-9 (Taylor & Francis Group)
ISBN 978-90-8585-314-5 (Wageningen University)

Preface

On a dark November night in 1975, José and I had a cosy dinner in a small restaurant at the Hertog Govertkade in Delft, opposite the pedestrian bridge. That night, we decided to follow one of our dreams, which was to work and live in countries less wealthy than ours. This decision certainly had far-reaching consequences. We enriched our lives in Surinam, Fiji, Kenya - where Renske was born - and Egypt, the birthplace of Jelle and Hilde. In the end, however, the price was high, maybe too high. José, your courage and determination have kept me going. This dissertation is for you.

Both the pedestrian bridge and the restaurant disappeared long ago, but the dream remained. Over the years, working in applied research and training allowed me to gain a wealth of new knowledge and experiences. This took time and, of course, enjoying life with Renske and Jelle had my priority. My colleagues at ILRI, especially Rien, always kept encouraging me about doing a PhD and, in the end, writing up the results has been quite rewarding.

In this dissertation, I discuss the role of subsurface drainage based on lessons learned in Egypt, India, and Pakistan. My experiences in Egypt and India, the two countries in which I have worked for about a third of my professional career, are at the core of the study. I could not have done this work without the help and support of so many colleagues, field staff, and farmers that it is impossible to mention everyone; please remember that all of you are in my heart. I especially want to acknowledge all of the farmers and their families. They allowed us to conduct our research in their fields. We asked them to implement a host of new concepts; we dug up their fields and asked them to modify their farming practices, all with no guaranty of success. Nevertheless, they had faith in our research activities and supported us. It gives me great satisfaction to realize that in most places I have worked, I would still be welcomed back as a friend. Clearly, these farmers deserve the credit.

I enjoyed the numerous discussions with Bart and Wouter. They helped me place my practical experiences in a scientific framework. Without their support I would likely still be in a field digging up drainpipes and looking into manholes for water. On the other hand, when I found myself too high up in a scientific cloud, Elizabeth was there, who in a very charming and stimulating way, always brought me back to reality. She symbolizes all women, who are not only much more practical, but also infinitely more romantic than men. I would also like to acknowledge the co-authors of the scientific papers on which this dissertation is based, since finding one's way through the multidisciplinary jungle is not always easy. Finally, I'd like to thank Bart and Wim for helping me assemble the pieces of my jigsaw puzzle by critically reviewing my synthesis.

Life was not always easy, and sometimes I buried myself in my work. It was Ymkje who helped me rediscover the joy of life and who, ironically, managed to stimulate me to complete my PhD thesis. Not in our prime time, of course; the days, nights, weekends and holidays we spent together - but in the remaining time, especially during my solitary missions abroad. And that I did, without having to sacrifice much, at least that is how I see it now.

Contents

Summary

The world's population is projected to grow from 6,500 million people today to more than 9,100 million in 2050. To feed this growing population and banish hunger from the world, food and feed production will need to be doubled. The majority of the increase has to come from investments in improved irrigation and drainage practices in existing agricultural areas as there is not much scope for horizontal expansion. At the same time, salinity affects 10 to 16% of all irrigated lands and the annual rate of land loss due to waterlogging and salinity is about 0.5 Mha per year. It is estimated that existing drainage systems in about 30 Mha of the irrigated areas will have to be replaced or modernized. Furthermore, in about another 30 Mha new drainage systems will have to be installed to overcome irrigation induced waterlogging and salinity problems. It is expected that about 50% of these systems will be subsurface drainage systems. At an average cost of € 1,250 per hectare this will require an investment of about € 19 billion or € 475 million annually over the next 40 years.

Scope of the study
In this research I have analysed the role of subsurface drainage in irrigated agriculture in arid and semi-arid regions and formulated recommendations for improving subsurface drainage practices. Drainage is not treated as a separate issue, but as part of Integrated Water Resources Management (IWRM). To enhance the role of subsurface drainage in IWRM, tools are presented for improving water efficiency and creating an enabling environment, highlighting the changing institutional roles and functions and the required management instruments. Based on lessons learned in the last 28 years in Egypt, India and Pakistan, supplemented when appropriate with experiences from other countries, the gradual change from a monodisciplinary towards a more multidisciplinary approach that integrates scientific, technical, socioeconomic and institutional elements is discussed. The thesis contains a synthesis that draws on a series of case studies published in international journals.

A modified layout of the subsurface drainage systems for rice areas
The first case study examines a modified layout of subsurface drainage systems in areas of Egypt with rice in the crop rotation. Tests were conducted on a modified design for the subsurface drainage system to reduce irrigation water losses from rice areas without restricting the subsurface drainage from 'dry-foot' crops. A three-step research programme was conducted over a six-year period. The principles of the modified design were studied under fully controlled conditions in experimental fields. Their applicability was studied in pilot areas under farmers' controlled conditions, followed by large-scale monitoring programmes to study whether the system would be accepted by the farmers and to assess the costs and benefits. The introduction of the modified layout of the subsurface drainage system in rice-growing areas in the Nile Delta resulted in savings in irrigation water of up to 30%. These benefits were obtained without any negative effect on either soil salinity or crop yield and with no increase in costs compared with the conventional system. The modified system protects crops other than rice from the damaging effects of these improper blocking practices, thus reducing maintenance.

Verification of drainage design criteria
The second case study presents the results of a monitoring programme to verify the design criteria of subsurface drainage systems in a 110 ha pilot area at Mashtul in the south-eastern part of the Nile Delta. The monitoring programme, which covered a 9-year period, showed that the crop yields increased significantly after installation of the subsurface drainage system. The increase was 10% for rice, 48% for *berseem* (Egyptian clover), 75% for maize and more than 130% for wheat. Part of these yield increases can be attributed to the decrease in soil salinity; the other part is the effect of improved water and air conditions in the root zone and improved agricultural inputs. The relation between crop yield and the depth of the water table showed that the optimum seasonal average depth of the water table midway between the drains is 0.80 m below the soil surface. The study confirms that that the current design criteria are still conservative: a better integration between irrigation and drainage will not only save irrigation water, but also reduce drainage discharges without sacrificing crop yields or increasing soil salinity.

Development of subsurface drainage strategies
The third case study presents the results of applied research studies that were set up to develop subsurface drainage strategies to combat waterlogging and salinity in five different agroclimatic subregions of India. Over a period of seven years, subsurface drainage systems were studied in six pilot areas in farmers' controlled fields, one experimental plot and one large-scale monitoring site. The study proves that, under the prevailing soils, agroclimatic conditions and social contexts, subsurface drainage by pipe or open drains is a technically feasible, cost-effective and socially acceptable technology for reclaiming waterlogged and saline land and sustaining agriculture in irrigation commands. Within one or two seasons after the installation of subsurface drainage systems, crop yields increased by an average of 54% for sugarcane, 64% for cotton, 69% for rice and 136% for wheat. These yield increases were obtained because water tables and soil salinity levels in the drained fields were respectively 25% and 50% lower than in the non-drained fields. Based on the research findings, drain depth/spacing combinations were recommended for various agroclimatic regions in India. The recommended drain depths (in the range 0.5 to 1.5 m) are significantly shallower than the depth traditionally recommended for the prevailing conditions in India (> 1.75 m). The corresponding shallower water tables avoid excessive drainage while at the same time effectively remove harmful salts brought in by the irrigation water.

The role of participatory modelling in research
The fourth case study discusses a participatory hydrological modelling approach to assess the off-site externalities caused by the disposal of drainage water. The study was conducted to develop an integrated approach to the restoration of the Kolleru-Upputeru wetland ecosystem on the east coast of Andhra Pradesh, India. The challenge was to overcome the hydrological and social complexities: the large variety of hydrological functions and the many stakeholders with different interests. In the approach, one of the main limitations of using simulation models, the lack of long-term data records, was tackled by matching the implicit (or tacit) knowledge of the stakeholders with the explicit knowledge brought in by the researchers. During a stakeholder workshop, the outcomes of the resulting problem analysis were matched with the stakeholders' views and experiences. Discussing model simulations with stakeholders proved to be a useful method of creating mutual understanding among the stakeholders about the complexity of the problems and that

single-issue solutions will not stop further degradation. As a result of the project, the stakeholders buried their differences and agreed on the outlines of an integrated approach.

Development of a drainage industry
The fifth case discusses how, over the last 50 years, subsurface drainage practices have changed from manual, small-scale installations to mechanized, large-scale systems. To keep up with the changing demands, new developments were needed in installation techniques, equipment and materials, as well as in the planning and organization of the implementation process. A shift from post-construction quality control to a total quality control system enabled high quality systems to be installed, even with the ever-increasing speed of installation. This was only possible because new modes for the implementation process were developed and implemented at the same time. Finally, it was shown that the introduction of modernized drainage machinery and installation techniques can only be successful if the people involved in the implementation process are properly trained. In addition to formal education and training, the 'in-service training' approach, based on the principle that the trainees go into the field instead of the classroom, proved to be a successful method.

The added value of research on drainage in irrigated agriculture
The sixth case study shows that applied research on drainage delivers value for money. Over the last forty years, countries like Egypt, India and Pakistan have invested heavily in applied research activities. These activities have helped to modernize subsurface drainage practices. Considerable savings have been achieved by the introduction of (i) new methods for investigating and identifying areas in need of drainage, (ii) new design and planning methods, (iii) new materials for pipe drains and envelopes, (iv) improved drainage machinery and equipment, and (v) improved installation, O&M methods and practices. Research has also helped to improve the organization of subsurface drainage operations and institutions. All these improvements could be achieved because these countries not only invested in research, but also in training all personnel to apply the new and innovative practices.

An integrated approach for capacity development in agricultural land drainage
The seventh case study presents lessons learned from using an integrated approach to capacity development in agricultural land drainage. Capacity development plays an essential role in improving irrigation and drainage practices in existing agricultural areas. Capacity development is a knowledge-creating process in which the more concrete or explicit aspects, such as training and institutional strengthening, are linked to local or tacit knowledge and aspects of ownership. Based on six examples from the Netherlands, Egypt, India, Pakistan, Indonesia and Malaysia, it is shown how research, training and advisory services can be linked into a knowledge-creating process. It is argued that these three elements (research, education and advisory services) have to be applied in an integrated manner. Research is required to link local knowledge with lessons learned elsewhere and serves to make knowledge explicit. Education is required to disseminate this explicit knowledge and, at the same time, to make the tacit knowledge of the participants explicit. Advisory services are needed to assist with the application of the newly acquired knowledge, thus completing the transformation from explicit to tacit knowledge. This approach is successful when those involved have the opportunity to go through the

knowledge-creating process several times. This will also lead to mutual trust between the cooperating partners, which is much enhanced when there is a long-term partnership.

These last two case studies show that countries do well to attach a research and capacity development component to large-scale implementation programmes.

The role of participatory research in project preparation
The eighth case study presents an example of how a participatory approach can enhance research. A participatory research study was conducted to improve the effectiveness of drainage in the Red River Delta in Vietnam. The stakeholders included not only the farmers, but also the other people living in the densely populated delta and their organizations. The study showed that both the physical and the institutional infrastructure of the drainage system constrains the performance of pumping stations. Only close cooperation between the stakeholders can improve the drainage in such complex systems like the polders in the Red River Delta. It is shown that improvement can only be achieved by a combination of technical and institutional measures. Although the study was conducted in Vietnam, in the humid tropics, the lessons learned with the participatory research are also applicable for research projects in the arid and semi-arid regions, where similar physical and institutional complex situations exist.

Synthesis: subsurface drainage practices in irrigated agriculture
In the synthesis, the challenges of making subsurface drainage work at a larger scale are addressed. The analysis of the subsurface drainage practices in Egypt, India and Pakistan shows that the installed systems are technically sound. They effectively prevent waterlogging and root zone salinity in irrigated land and consequently increase crop yields and rural income. The research supports the prevailing view that deep drains are unnecessary for salinity control and that better options for operational management can further reduce drain depths and design discharges. A better integration of irrigation and drainage will help to save irrigation water and further reduce drain discharges without sacrificing crop yields. The introduction of new types of installation equipment and materials and the corresponding implementation practices has made large-scale implementation feasible. The economic analysis shows that these subsurface drainage systems are a very cost-effective measure for combating waterlogging and salinity in irrigated agriculture. The recent rise in the price of the major food commodity prices will increase the economic returns even further.

However, although it can be concluded that the installed systems are technically sound and cost-effective, drainage development lags behind irrigation development and consequently a substantial part of the irrigated areas suffer from waterlogging and salinity. An exception is Egypt, where the government took full responsibility for the implementation of subsurface drainage systems. But even in Egypt, handing over O&M to the farmers is problematic. This is mainly because the subsurface drainage practices are designed and implemented by government, with the users, the small farmers, having little responsibility and making little input: a top-down approach in which location-specific conditions and farmers preferences are hardly taken into consideration. Furthermore, the emphasis has been more on the technical aspects (the physical infrastructure), while the organizational aspects (institutional infrastructure) have been largely neglected. Although most farmers are poor and do not have the means to invest in subsurface drainage, they clearly see the benefits and are willing to contribute.

The way forward: enhancing the role of subsurface drainage
To reverse the negative trend in salt build-up and waterlogging, I have identified the following challenges for enhancing the role of subsurface drainage: (i) balancing top-down against bottom-up, (ii) from standardization to flexibility, and (iii) focus on capacity development.

(i)   *Balancing top-down against bottom-up.* Participation by farmers needs to be increased in all phases of the implementation process. More attention needs to be paid to the identification of the stakeholders and their needs, preferences and willingness to contribute. A participatory learning and action approach is an effective and efficient method for assessing the need for drainage, creating a mutual understanding of the problems and developing an integrated approach to development. Participatory modelling is a useful tool for creating a better understanding among the stakeholders of the complexity of the problems and the effectiveness of solutions. Through these participatory tools, the link between technical aspects (requiring physical solutions) and organizational aspects (requiring institutional changes) can be enhanced.

(ii)   *From standardization to flexibility.* Instead of standardized design and implementation practices, a much more flexible approach based on location-specific conditions and stakeholders' preferences is recommended. Integration between the irrigation and drainage network needs to be improved. The challenge is to find a balance between the individual need for drainage, which varies from field to field, and the fact that drainage at farm level is a collective activity. This requires better operational control. Controlled drainage will allow the farmers to optimize their on-farm water management, based on the specific conditions and their own preferences. Furthermore, it enables the farmers to respond to changes in land use and/or the effects of climate change.

(iii)   *Focus on capacity development.* More stakeholder participation and more flexibility can only be achieved if the tacit knowledge of these stakeholders is linked to the explicit knowledge of researchers, planners and designers. In this knowledge-creating process, the explicit knowledge of the researchers can be internalized (learning) through education and training and then linked to the tacit knowledge of the stakeholders by socialization (sharing experiences) through research. Bringing tacit and explicit knowledge together yields new knowledge through externalization. In turn, this can again be combined with explicit knowledge from elsewhere (synthesis) and used in guidelines and by advisory services. In this process, the four steps of the knowledge-creating process may provide a useful framework for designing a capacity-building strategy.

I am convinced that the above recommendations will facilitate the further introduction of subsurface drainage in arid and semi-arid irrigated areas throughout the world, and through this contribute to a better, more sustainable use of the precious land and water resources in these areas. This requires policy and institutional changes for which governments would have to take the lead. In consultation with the stakeholders, governments need to develop a drainage policy that emphasizes the need to treat the reclamation of waterlogged and salt-affected areas in irrigation projects and the creation of fresh irrigation potential or its utilization with equal importance. This policy would have to include a time-bound action plan to safeguard these irrigated lands against these problems. Farmers would have to be willing to participate and pay part of the cost, but as the benefits often go beyond the direct interest of the farmers concerned, governments need to finance or prefinance part of the costs. The stakeholders, including the farmers, need to contribute, either in cash or in kind (labour).

Further research and development is needed to meet the specific needs of emerging and developing countries, which each have their own specific climatic, physical and social conditions, and to cope with climate change, land use changes and requirements related to the quantity and quality of drainage water. These changes will require modifications in subsurface drainage practices: from planning and design to implementation, O&M. It is the farmer who has to adapt his farming system to these changing needs. The challenge for the research and education community is to support farmers in managing their fields in a more sustainable way and to enable them to cope with these changes. Only if these challenges are met will investments in irrigated agriculture in arid and semi-arid areas be protected, increasing its sustainability and its chances of feeding the growing world population.

1 Introduction

1.1 Rationale of the study

The world's population is projected to grow from 6,700 million people today to more than 9,200 million in 2050 [258]. At present, 2,300 million people are directly or indirectly dependent on agriculture and 75% of the 900 million of the worlds' poor people live in rural areas: 80% in Africa, 85% in South Asia, 63% in East Asia and 47% in Latin America and the Caribbean [300]. Increased productivity, profitability and sustainability of small-scale agriculture are essential to reduce poverty in rural areas and thus important for achieving the Millennium Development Goals (MDG), in particular no. 1 Eradicate extreme poverty and hunger and no. 7 Ensure environmental sustainability [257; 300]. To be able to feed the growing world population and to banish hunger from the world, food and feed production will need to be doubled in the coming 25 to 50 years [145]. At the same time, the growth rates of yields of major crops in developing countries are declining (Figure 1.1). This decline in annual growth rates is not necessarily related to a decline in absolute yield growth per annum. An important explanation for the decrease in these growth rates might be the declining public spending on agricultural research and development over the last thirty years [300]. In 2006, the World Bank argued that it is time to increase levels of investments in agricultural water management as an engine of growth [298] .

At present, about 272 Mha (or 18%) of the arable and permanent cropped areas are irrigated (Table 1.1), contributing as much as 35–40% of the gross agricultural output. There is not much scope for increasing the irrigated area (Figure 1.2), thus the majority of the increase in irrigated agriculture will have to come from investments in improved irrigation and drainage practices in existing agricultural areas [218].

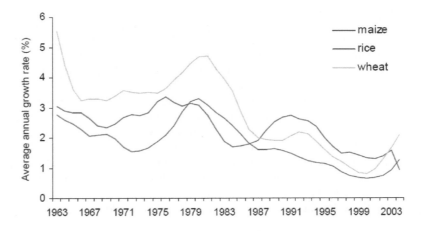

Figure 1.1   Average annual growth rate in yield of selected crops in developing countries [300]

Table 1.1   Key indicators of the agricultural sector [105]

Indicator	Unit	World	Egypt	India	Pakistan
Total geographical area (TGA)	(Mha)	13,425	100	329	80
Arable & permanent cropped area (APC)	(Mha)	1,497	3.4	170	22
Population	(Million)	6,134	71	1,050	150
Population in agriculture	(Million)	3,211	40	755	99
Population in agriculture	(%)	52	56	72	66
Population density with ref. to TGA	(No.km^{-2})	45	70	319	188
Population density with ref. to APC	(No.km^{-2})	410	2,074	617	678
Food production (cereals)	(MT)	2,086	19	232	28
Productivity for cereals	(kg/ha)		7,249	2,356	2,302
Gross national income per capita	(US$)		1,390	540	520
Irrigated area	(Mha)	272	3.4	57.2	16.7[a]
Irrigated area % of APC	(%)	18	100	34	80
Drained area	(Mha)	190	3.0	2.5	7.5
of which equipped subsurface drainage	(Mha)		1.9	0.025	0.32
Salt-affected areas	(Mha)		1.0[a]	6.7[a]	2.4[a]
of which also waterlogged	(Mha)		0.6[a]	4.5[a]	1.7[a]

[a]   data Egypt [7] , data India [86], data Pakistan [301]

At the same time, salinity problems are a fact of life to irrigation in arid and semi-arid regions. Under the prevailing dry and high evaporation conditions, salt concentrations and river depletion are two inevitability collaterals of irrigated crop production in these regions [236]. Salinity and the related waterlogging problems affects about 10–16% of the irrigated lands [219; 249]. In Asia this figure is nearly 40% [300]. Worldwide, the annual rate of land loss due to waterlogging and salinity is about 0.5 Mha per year [232]. The history of ancient Mesopotamian illustrates that salinization when not properly recognized and treated can be a time bomb waiting to explode upon the agricultural scene [176].

Drainage is an essential tool to combat waterlogging and salinity. At present, however, only about 190 Mha, or 13% of the world's arable land, is provided with some sort of drainage [105]. Drainage development is mainly driven by the level of agricultural development and the related technical merits and (farm) economic viability as an instrument for more profitable land use and further agricultural development [235]. Two-thirds of the drainage systems (about 130 Mha) have been installed in rainfed areas in the humid regions and one-third (about 60 Mha) in irrigated areas in semi-arid and arid regions [219]. Most of these drainage systems are at least 30–40 years old.

Index: 1970 = 100

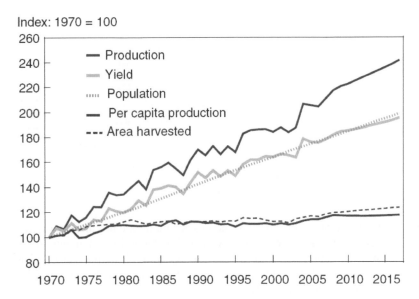

Figure 1.2   Development of world grain and oilseed production, yield, area harvested, population and per capita production [253].

It is estimated that, in the irrigated areas, existing systems will have to be replaced or modernized in about 30 Mha. Furthermore, in about another 30 Mha new systems will have to be installed to overcome irrigation-induced waterlogging and salinity problems. It is expected that about 50% of these systems will be subsurface drainage systems. At an average cost of € 1,250/ha [1], this will require an investment of about € 19 billion or € 475 million annually over the next 40 years [150].

Drainage plays an important role in agricultural and rural development in many countries and is one of the pillars for sustaining world food production (Figure 1.3). Drainage at field level can be divided into surface and subsurface drainage. Surface drainage is the diversion or orderly removal of excess water from the surface of the land. Subsurface (or horizontal) drainage [2] is the removal of excess water and salts from soils via groundwater flow to the drains [192]. Tubewell or vertical drainage is a special type of subsurface drainage for controlling the water table through a group of adequately spaced wells. Subsurface drainage has been practised for thousands of years; large-scale introduction, however, only started around the middle of the last century, when the prevailing empirical knowledge of drainage and salinity control gained a solid theoretical foundation. According to Bos and Boers [42] *'sound theories now form the basis of modern drainage systems, but there will always remain an element of art in land drainage. It is not possible to give beforehand a clear-cut theoretical solution for each and every drainage problem: sound engineering judgement on the spot is still needed, and will remain so'*.

---

[1] All prices in this thesis have been converted to euro (€1.00 = US$1.35) at 2007 prices, except when stated otherwise.

[2] In the literature subsurface drainage is also referred to as 'pipe' and 'tile' drainage.

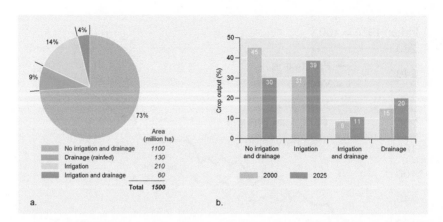

Figure 1.3   Drainage plays an essential role in sustaining food production: (a) Worldwide
             agricultural areas equipped with and without irrigation and /or drainage
             systems, and; (b) Percentages of agricultural output (crop yield) from
             agricultural land with and without irrigation and drainage facilities now
             (2000) and in 2025 [105].

The installation practices evolved from purely manual installation on individual farm plots
to fully mechanized installation programmes covering thousands of hectares. To make this
rapid change possible, research was needed to modernize subsurface drainage practices and
staff had to be trained in these modernized drainage machinery and installation techniques,
as well as in the planning and organization of the implementation process. These
developments are still continuing to meet the specific needs of installation in the emerging
and least developed countries, under climatic, physical and social conditions that differ
from the ones for which they have been designed. Furthermore, the specific needs of
drainage are also changing, particularly with regard to the quality of drainage water, and
these require changes in drainage system design and corresponding installation practices.

The role of drainage varies between the different agroclimatic zones [219]. In the
temperate zone, mainly located in the northern hemisphere, the role of drainage is to
prevent waterlogging by removing excess surface and subsurface water resulting from
excess rainfall. In the arid and semi-arid zone, the role of drainage is to prevent irrigation-
induced waterlogging and salinity, not only by removing excess surface and subsurface
water but also by removing soluble salt brought in by the irrigation water. In the humid and
semi-humid zone, the role of drainage is to prevent waterlogging and salinization to
various degrees. About 64% of the drainage is located in the temperate zone, 24% in the
arid and semi-arid zone and 12% in the humid and semi-humid zone.

In the arid and semi-arid zone, irrigation is a tool that allows farmers to cope with
inadequate and unreliable rainfall [239] and drainage is a tool for controlling the water
table and root zone salinity. In irrigated agriculture, integration of irrigation and drainage is
needed (i) to manage the water balance in order to reduce water requirements for irrigated
agriculture, (ii) to manage the salt balance, because even with improved irrigation practices
the leaching requirements have to be met, and (iii) to manage the financial balance in order
to reduce the public costs of operating irrigation and drainage schemes [238]. Countries

like Egypt, India and Pakistan have invested heavily in irrigation and, to protect these investments and to increase the sustainability of their agricultural lands, in drainage. In these countries, the majority of the population is still employed in the agricultural sector and the majority of these farmers are smallholders, owning less than 1 hectare of cultivable land. About one-fifth of the drainage systems installed in irrigated lands are in these three countries (Table 1.1). To overcome the current waterlogging and salinity problems, more areas will have to be drained. In the 1960s the Egyptian Government started an ambitious programme to provide all agricultural land in Egypt with drainage by 2012 [150]. In Pakistan 9.1 Mha of agricultural land still needs to be provided with drainage [301] and the same applies to the 6.7 Mha of waterlogged and salt-affected lands in India.

Despite these needs, the role of drainage in irrigated agriculture in most of the global, national and regional agricultural and rural development agendas and programmes has been quite insignificant in the last decade [6]. In many 'vision' and policy documents, irrigation is seen as the key developmental intervention, while drainage is only mentioned as a necessary preventive/remedial supplementary measure to irrigation, but not as a development instrument in its own right [145; 209]. Even in the International Commission on Irrigation and Drainage (ICID), the most important association of professionals in the field of irrigation and drainage, drainage has been given only limited attention. None of the Commission's congresses have had a dedicated drainage focus and just 5–10% of the papers presented at most of them dealt with drainage [233]. The international drainage workshops – the first of which was organized by the International Institute for Land Reclamation and Improvement (ILRI) in Wageningen in 1979 and the 10th and most recent held in Helsinki/Tallinn in 2008 – have not been able to change this attitude. This narrow view of drainage, which is also held in many of the international development agencies, is especially remarkable in view of the important contribution that drainage has made to agricultural and rural development in numerous developed countries [163; 232].

Fortunately, the tide is turning and agricultural water management and drainage are slowly coming back into the limelight. For example, the president of ICID has given drainage a prominent place in his top ten irrigation technology issues [129]. He stated that '*drainage must not be forgotten*' and '*that we must not forget that drainage can improve and sustain production in rather more parts of the world than irrigation on its own*'. Other intergovernmental platforms also argue for new directions in agricultural knowledge, science and technology to address issues such as how to provide safe water, maintain biodiversity, sustain the natural resource base and minimize the adverse impacts of agricultural activities on people and the environment [104]. The introduction of new crop varieties and crop diversification will require improved water management practices, including drainage. Improved drainage practices are also needed because the hydrological environment is also changing: in many regions less water is available due to increasing scarcity and upstream use. In other regions, water quality is deteriorating or the discharge of drainage effluent is restricted. Furthermore, one of the effects of climate change may be that extreme weather conditions will occur more often: both wet and dry periods may be on the increase [268]. Finally, socioeconomic conditions are also changing and participatory approaches are required in response [248]. Above all, however, adequate waterlogging and salinity control at farm level, including maintenance, is an institutional challenge [215] .

Drainage, however, should not be treated as a separate issue: drainage is part of Integrated Water Resource Management (IWRM) [9]. IWRM is emerging as an alternative to the top-down approach that was central to the water resources management in the 20th century [48]. IWRM is a process that promotes the coordinated development and management of water, land and related resources in order to maximize economic and social welfare in an equitable manner without compromising the sustainability of vital ecosystems [83]. Operationally, IWRM approaches involve applying knowledge from various disciplines as well as the insights from diverse stakeholders to devise and implement efficient, equitable and sustainable solutions to water and development problems. IWRM draws its inspiration from the Dublin principles [107]. It requires a participatory approach, emphasising the need for more stakeholder involvement, both male and female, including their role as decision makers and water users. An IWRM approach is an open, flexible process that brings stakeholders together to make sound, balanced decisions in response to specific water-related challenges. Thus, IWRM represents a major challenge for policymakers. It requires a break with tradition, from the sectoral to integrated management; from top-down to stakeholder demand responsive approaches; from supply fix to demand management; from command and control to more-co-operative or distributive forms of governance and from closed expert driven management organizations to more open, transparent and communicative bodies. The IWRM toolbox contains three groups of tools to reach these objectives: (i) the enabling environment, (ii) institutional roles, and (iii) management instruments [83]. Participation and capacity development are also key elements of IWRM. Participation by relevant stakeholders is required so that they can manage the issues together because, typically, no-one has all the necessary legal, financial and other resources to do this satisfactorily on their own [190]. Capacity development aims to develop institutions, their managerial systems and their human resources to make the sector more effective in delivering services [256]. It therefore addresses three levels – the individual, the institution and the enabling environment – and is essential because stakeholders need to learn about and recognize each other's concerns and viewpoints. They need to arrive at a shared understanding of the issues at stake and of possible solutions. Although it is generally recognized that drainage is an essential part of IWRM practices, and in some case this role has been successfully demonstrated [85], in general the role drainage plays has not been worked out in detail.

1.2 Scope of the study

The role of subsurface drainage in arid and semi-arid areas is discussed to highlight why drainage is needed to contribute to the increasing demand for food, to safeguard investments in irrigated and non-irrigated agriculture and to conserve land resources. Drainage is treated as an integrated part of water management. We can draw an analogy with the human body. Without the removal of excess water and the hazardous elements dissolved in it from our body, we would die within one week from poisoning. Leaching of these hazardous elements is an absolute requirement for survival. Drainage has the same functions as our kidneys: it serves to discharge excess water that is received through precipitation, surface runoff from upstream areas and irrigation. Furthermore, drainage removes salts, imported by irrigation water, from the root zone.

To put the role of subsurface drainage in IWRM into practice, tools for improving agricultural water management and creating an enabling environment are presented, highlighting the changing institutional roles and functions and the required management instruments. Based on lessons learned in the last 25 years in many projects in a number of countries in the Middle East, Africa and South and South-East Asia, the gradual change from a monodisciplinary towards a more multidisciplinary approach, integrating scientific, technical, socioeconomic and institutional elements, is discussed.

1.3 Objectives

The overall objective of the study is to highlight the potential of subsurface drainage for improving irrigated agriculture in arid and semi-arid regions, and in particular:
- to review subsurface drainage practices in irrigated agriculture in Egypt, India and Pakistan;
- to assess the performance of these subsurface drainage systems, especially in relation to IWRM;
- to identify improvement options in the planning, design, installation, O&M (O&M) practices in order to increase efficiency, equity and environmental sustainability;
- to show when these improvements are useful in contributing to the increasing demand for food, in safeguarding investments in irrigated and non-irrigated agriculture, and in conserving land resources.

Based on the objectives, the following research questions were formulated:
- why is subsurface drainage an accepted practice in Egypt (where it is implemented in almost all irrigated lands) and not in India and Pakistan (where as a consequence millions of hectares are waterlogged and/or salinized)?
- how can the integration between irrigation and drainage be improved?
- under what conditions is subsurface drainage a technically feasible, cost-effective and socially acceptable option for sustaining agriculture in irrigated lands?
- what are the main challenges in making subsurface drainage work at a larger scale?

1.4 Hypothesis

The hypothesis of this study is that four constraints hamper the large-scale introduction of subsurface drainage in irrigated areas, in particular in emerging and least developed countries:
- farmers, although they clearly see the benefits of drainage, are in general too poor to pay the full cost of drainage, but they need to pay the sustainability cost[3];
- subsurface drainage can only be successful in controlling salinity if sufficient good quality irrigation water or monsoon rainfall is available for leaching.

---

[3] The sustainability cost includes all operation, maintenance and renewal costs, including all the staff costs linked to the service, but does not include the full financial cost of the initial investment or of past upgrading [250].

Supplementary measures in soil and water management, for example the application of gypsum, the introduction of salt-tolerant varieties and irrigation efficiency improvement, are needed to enhance the positive effects of subsurface drainage. Thus a trade-off between the additional investments in soil and water management and savings in drainage costs would have to be considered;

- drainage in irrigated agriculture has always had a lower priority than irrigation. For farmers, irrigation is needed today ('no water, no crop'), but drainage is a more preventive measure as salinity is a slow process threatening sustainability;
- drainage at farm level, even more than irrigation, is a collective activity. Appropriate institutional arrangements for farmers' participation and organization have to be developed.

1.5 Methodology

The role of subsurface drainage is analysed based on lessons learned in Egypt, India and Pakistan. These lessons are derived from my working experience, supplemented, where appropriate, by a literature review. After I obtained my MSc in Civil Engineering at Delft University of Technology in 1980, I worked in drainage-related water management projects in Fiji, Kenya and Egypt. In 1989, I joined the International Institute for Land Reclamation and Improvement (ILRI), which merged with Alterra in 2001, and continued working in research, training and consultancy. Over the years, my work gradually changed from a monodisciplinary towards a more multidisciplinary approach, integrating scientific, technical, socioeconomic and institutional elements. My experiences in Egypt and India, the two countries where I have worked for about one-third of my professional career, are at the core of the study. These experiences have been documented in numerous reports, papers and books. The lessons learned in Egypt and India have been complemented by lessons learned in Pakistan, a country I visited only once, but that I know well from the literature. Where appropriate, these lessons have been supplemented by experiences from the other countries where I worked and from the literature. I have analysed the role of subsurface drainage based on the four main phases of subsurface drainage practices: planning, design, installation and O&M. I regard drainage as a part of IWRM and present tools designed to increase water efficiency, create an enabling environment and change institutional roles and functions.

1.6 Benefits of the research

The study presents recommendations on how to improve subsurface drainage practices in arid and semi-arid regions. Professionals can use these recommendations to improve planning, design, implementation, and O&M practices. The recommendations can also help planners and decision makers to address issues related to an enabling environment (who will pay the costs of drainage?) and the changing roles of institutions to improve farmers' participation and organization.

1.7   Outline of this thesis

This PhD thesis contains a synthesis based on a series of case studies published as papers in peer reviewed international journals. For the literature review, an additional 76 cases covering subsurface drainage practices in Egypt, India and Pakistan were analysed [194].

In Chapter 2, the subsurface drainage practices in Egypt are discussed. The chapter includes two case studies. The first case study examines a modified layout of subsurface drainage systems in areas with rice in the crop rotation. The principles of the modified layout have been investigated in experimental fields under fully controlled conditions as well as in farmers' controlled fields. In addition, the performance of the modified layout was monitored in two areas of around 20,000 ha each. The second case presents the results of a monitoring programme to verify the design criteria of subsurface drainage systems. This monitoring programme, which covered a 9-year period, was conducted in Mashtul in a 110 ha pilot area in the south-eastern part of the Nile Delta.

In Chapter 3, the subsurface drainage practices in India are discussed. This chapter also includes two case studies. The first case study presents the results of applied research studies that were set up to develop subsurface drainage strategies to combat waterlogging and salinity in five different agroclimatic subregions of India. Subsurface drainage systems were studied in six pilot areas in farmers' controlled fields, one experimental plot and one large-scale monitoring site. The second case study discusses a participative modelling study that was conducted to develop an integrated approach to assess the off-site externalities caused by the disposal of, among others, drainage water in the Kolleru-Upputeru wetland ecosystem on the east coast of Andhra Pradesh. A four-step participative modelling approach was developed to bring the stakeholders together, to create a mutual understanding of the need for an integrated approach, instead of taking single-issue measures, and to agree on follow-up steps needed to sustain both the livelihood of the people as well as the Kolleru and Upputeru ecosystem.

In Chapter 4, the subsurface drainage practices in Pakistan are discussed. This chapter is based on a literature review in which 25 cases covering the four main phases of subsurface drainage practices – planning and organization, design, installation, and O&M – were analysed.

In Chapter 5, two case studies are presented to show how subsurface drainage practices have changed over the last 50 years and the role research has played to improve these practices. The first case discusses how subsurface drainage practices have changed from manual installation to large-scale implementation. It shows that to keep up with the changing demands, new developments were needed in installation techniques, equipment and materials, as well as in the planning and organization of the implementation process. The second case shows that applied research on drainage delivers value for money. It shows that research findings have helped to modernize subsurface drainage practices and that considerable savings can be achieved by introducing new methods for investigation, design, planning, installation (including new materials and equipment) and O&M.

In Chapter 6, two case studies are presented to illustrate the role capacity development plays in improving subsurface drainage practices. In the first case, lessons learned with an integrated approach for capacity development in agricultural land drainage are presented. This approach is based on three elements: research, education and advisory services. It is argued that capacity development is as much a process as a product in which these three elements have to be applied in an integrated manner. In the second case, an example is presented of how participatory capacity and empowerment of stakeholders can be enhanced. It discusses a participatory research study to improve the effectiveness of drainage in the Red River Delta in Vietnam. Besides technical innovations, recommendations to reform the complex institutional setting were formulated. Although the study was conducted in the humid tropics, the lessons learned with the participatory research are also applicable for research projects in the arid and semi-arid regions, where similar physical and institutional complex situations occur.

In Chapter 7, a synthesis of subsurface drainage practices in arid and semi-arid areas is presented. The question whether subsurface drainage is a technically feasible, cost-effective and socially acceptable technology is addressed. I discuss how the integration between irrigation and drainage can be improved and address the challenges to make subsurface drainage work at a larger scale. These challenges include institutional and policy issues in transforming the present top-down approach into a more participatory approach, the drainage system requirements, including the need for flexibility and control, methods to achieve this more participatory approach and the capacity development needed to make these changes successful.

Finally, in Chapter 8 I have identified challenges for enhancing the role of subsurface drainage to increase the sustainability of irrigated agriculture in arid and semi-arid regions.

2   Subsurface drainage practices in Egypt

2.1   History of irrigation and drainage in Egypt

The Nile River Basin is, like the Indus basin, one of the oldest agricultural areas in the world [23]. As Egypt's average annual rainfall ranges from 1.5 mm in the south (near Aswan, about 900 km south of Cairo) to 150 mm in the north (in the coastal regions bordering the Mediterranean Sea, about 150 km north of Cairo), agriculture has always depended upon irrigation. The River Nile represents the only renewable source of water for Egypt's 3.4 Mha agricultural lands (Figure 2.1).

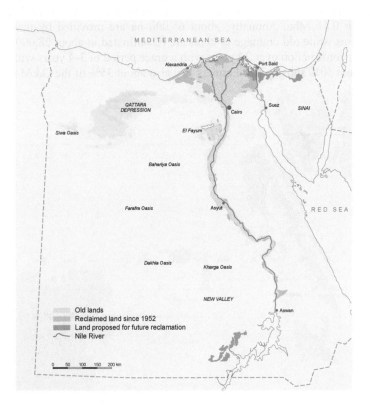

Figure 2.1   Agriculture in Egypt depends entirely on irrigation from the River Nile

Since the days of the Pharaohs until the 19th century, basin irrigation has been practiced. For this ancient method of irrigation, based on the natural regime of the Nile, the natural drainage capacity of the land was sufficient to protect the area against the twin problem of waterlogging and salinity. In the 19th century, new crops, i.e. cotton and sugarcane, were introduced that required water when the Nile's water levels were low. This resulted in the construction of barrages in the River Nile and a network of irrigation canals and open drains. The completion of the Aswan High Dam in 1968 finally eliminated the Nile's season floods and allowed all agricultural lands to be brought under perennial irrigation. Nowadays, the main crops are cotton, sugarcane and paddy in summer and wheat and *berseem* (Egyptian clover) in winter. Land holdings are small with about 78% less than 2.0 ha [7]. The elimination of the seasonal fluctuation in the River Nile resulted in higher piezometric pressure in the aquifer underlying the agricultural areas and thus reduced the natural drainage capacity (Figure 2.2). Together with the increased percolation from irrigation this gradually resulted in waterlogging and salinity problems in large areas. The open drainage systems, constructed since the second part of the 19th century, were not sufficient to overcome these problems and in the 1960s the Egyptian Government embarked upon an ambitious programme to install subsurface drainage systems in all agricultural lands by 2011 [75]. Currently, main drainage systems have been improved in about 2.4 Mha of which 1.9 Mha have been provided by subsurface drainage [7]. On top of this, the upgrading of subsurface drainage systems older than 30 years has been initiated, covering about 0.41 Mha. Annually, about 65,000 ha are provided by new subsurface drainage systems while old drainage systems are rehabilitated in about 28,000 ha. Farmers pay the cost of construction over 20 years with a grace period of 3-4 years without interest, this means about 50% subsidy [6]. Farmers pay also about 35% of the O&M through land taxes.

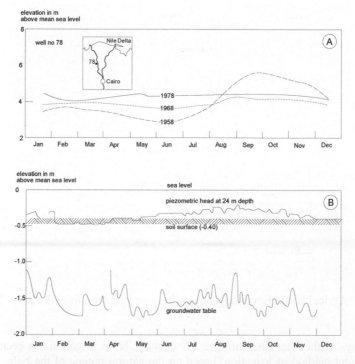

Figure 2.2   A: Fluctuation of the piezometric head in the Nile Delta Aquifer, before (1958) and after (1978) the construction of the Aswan High Dam (completed in 1967), and B: Piezometric head and the fluctuation of the water table in Sherashra pilot area [295].

2.2   Organization of the drainage sector

Egyptian Public Authority for Drainage Projects

To implement this ambitious drainage programme several institutions were created within the Ministry of Water Resources and Irrigation (MWRI). In 1973, the Egyptian Public Authority for Drainage Projects (EPADP) was established [75]. EPADP is a semi-autonomous authority, headed by a Chairman with the rank of First Under-Secretary directly responsible to the Minister of Public Works and Irrigation. EPADP has one Vice-Chairman supported by five regional Departments, each headed by an Under-Secretary. At present EPADP employs about 4,000 permanent staff at its headquarters and directorates and about 3,000 casual labourers who mainly work in the maintenance of drainage systems. EPADP is responsible for the field drainage works, including the planning of projects, data collection, preparation of designs, contracting and supervising the installation of subsurface drains, monitoring of the impact of drainage, budgeting, and operating project accounts. Nowadays, the implementation is done by public and private contractors, but EPADP still has comprehensive responsibility for the other activities. In addition, EPADP is responsible for the remodelling of the main drainage system, including pumping stations and, since 1992, also for the maintenance of all open drains.

Drainage Research Institute

To assist EPADP with this programme, a new research infrastructure was set-up within MWRI. The Drainage Research Institute (DRI) was established in 1976 as part of the National Water Research Centre (NWRC) of MWRI to conduct applied research, monitoring, testing, and evaluation of drainage methodologies and techniques. Its activities are intended to support EPADP's implementation programme and to solve their technical problems. DRI employs about 72 professional staff and 150 supporting and administrative staff. Since its establishment, DRI has cooperated with Alterra-ILRI through a number of bilateral projects. The first project (1976-1979) established the Egyptian-Dutch Advisory Panel on Land Drainage, with various drainage research and capacity building components [70; 265]. It was followed by a series of bi-lateral technical assistance projects. In the first phase of this long-term cooperation, the emphasis was on technical cooperation [64]. In the follow-up projects, the emphasis was on transforming DRI in a robust research organization [63].

Other institutes involved in drainage research

The Research Institute for Ground Water (RIGW), another institute of the NWRC, carries out groundwater surveys and groundwater development studies [150]. This institute also provides the drainage implementation programme with significant research input. It has investigated the seepage from the new land schemes located at higher elevations, which has caused waterlogging and salinization problems in the old lands. RIGW has implemented studies on the technical and economic feasibility of vertical drainage in these zones, known as the fringe zones of the Nile Valley.

The Soils, Water and Environment Research Institute (SWERI) of the Ministry of Agriculture and Land Reclamation (MALR) conducts soil surveys on irrigated land. SWERI has conducted extensive research on the drainage of heavy clay soils in the northern part of the Middle Delta. SWERI has also undertaken research on concurrent applications of gypsum and sub-soiling and its effect on drainage enhancement.

Collector-user associations and water boards
For maintenance purposes, more than 2,000 collector-user associations per collector drain (about 100 to 300 ha) were established on a voluntary basis [296]. They had no legal or institutional framework and failed because they were given too little to do. Since 1995, some 50 elected water boards have been established at secondary canal command level (500 to 750 ha). A discussion is going on to upscale them to district level (10,000 to 15,000 ha) to get better integration between irrigation and drainage.

2.3   Planning of drainage projects

EPADP's Planning Department is responsible for setting up the five-year and annual execution plans, along with the financing of projects [76; 141]. Negotiations with financiers of EPADP projects are done through this Department. A key element in the planning is the policy to carry out projects in clusters or land blocks, which at present are around 3,500 to 8,500 ha in size. About 95,000 ha of subsurface drainage systems are installed each year. This requires a strict and well-balanced project preparation and planning, which was developed and modified over the years. The preparation and planning cycle includes three steps: (i) identification and planning, (ii) investigation and design, and (iii) tendering and contracting [22; 75].

Identification and planning includes four steps. In the Identification stage the type of the drainage problem is identified on the basis of available information, augmented by minor analysis. In the Pre-feasibility stage a reconnaissance survey is made to make a preliminary diagnosis of the problem and a rough outline of possible solutions. In the Feasibility stage all relevant information is collected through a semi-detailed type of field investigations (map scale: 1: 10,000, 1: 25,000) and a final solution is chosen. In the Final stage detailed field investigations are undertaken and detailed plans are prepared to serve as working documents for implementation, i.e. detailed designs and construction drawings, specifications and planning of the execution.

Investigation and design begins by obtaining surveying maps of the project area from the Egyptian Survey Authority, with updated information on villages, towns and built-up structures. Following the preparation of project maps, the field investigation programme is prepared for site sampling locations (500 x 500 m grid). Groundwater levels, soil permeability and salinity are measured and soil samples are collected and sent to DRI for analysis. Based on the soil permeability and groundwater levels, the layout of the subsurface drainage system is prepared and longitudinal profiles of the collectors are made.

Tendering and contracting starts after the design album and the lists of quantities have been prepared. The project is tendered among pre-qualified drainage contractors. Local public and private sector contractors do the earthwork for remodelling open drains and installing subsurface drains. Structures to be rebuilt in open drains are awarded to local contractors in the private and public sectors, following local tendering procedures.

Operational research

Accurate data on capacities, efficiencies, availability of machines, equipment and contractors are needed for the planning and contracting of the drainage projects. To collect such data, an Operational Research Unit (ORU) was established in 1993 within the Planning and Follow-up Department (PFD) of EPADP [141]. ORU conducted a number of operational research activities, i.e. an inventory of all drainage machine working all over Egypt was made to quantify the machine specifications, the project-related data and the performance. Time and motion studies were conducted to quantify the effective working time of the machines and calculate the capacity of the field and collector machines. Some salient results of these studies were [142]:

- 59% of the field drainage machines and 76% of the collector drainage machines were operational;
- a good relation could be established between the performance of the machine and its age. The performance of the machines was classified as 'good', 'moderate', 'bad' or 'beyond repair'. Both field and collector drain machines were in a 'good' condition up to the age of approximately 7 years, changing to a 'moderate' condition between the age of 8 and 15 years. After approximately 16 years, the condition between field and collector drainage machines started to deviate. Of the field drainage machines older than 16 years nearly 75% were 'beyond repair' and 14% were in a 'bad' condition. Figures for collector drainage machines were 13% and 43%, respectively. Thus, collector machines have a longer lifespan than field drainage machines. But, as efficiency increases over the years the operational lifetime will drop to 10 to 12 years in the future;
- the effective time is about 198 working days per year and 3 and 4 hours per day for respectively collector and field drainage machines;
- the installation capacity of collector drainage machines decreases from 100 m/h for new machines to 55 m/h for machines older than 15 years. For field drainage machines, the figures are respectively 380 and 90 m/h.

The results are used to improve the planning of the execution of the drainage projects, both in time and manpower and also to select supplies of new machines as they can be assessed on the performance of machines bought in the past.

2.4 Design principles

Layout

The subsurface drainage system installed in Egypt consists of subsurface field (named laterals in Egypt) and collector pipes that run by gravity [75]. The buried pipes form a regular pattern of field and collected drains (Figure 2.3). The piped collectors discharge into open main drains from where the drainage water is pumped into large open outfall drains which eventually discharge into the River Nile or the sea. Pumping is necessary almost everywhere in the Delta and the Valley, except in some areas in Upper Egypt, where there is enough gradient to dispose of the effluent freely by gravity. To reduce water losses from areas cultivated with rice without restricting drainage from other areas, a modified layout of the subsurface drainage system has been developed (Section 2.7).

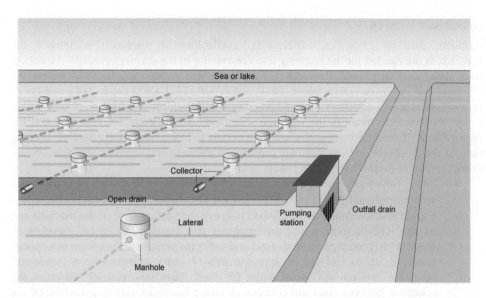

Figure 2.3   Schematic representation of the subsurface drainage system used in Egypt

Design criteria

The design criteria of the subsurface drainage system are based on the requirements of the most critical crop, which is considered to be cotton [4]. The design is based on average hydrological conditions. The design criteria are divided in agricultural and technical criteria [155]. The agricultural criteria are based on the effect of land drainage on crop production under the prevailing agricultural and hydrological conditions. The technical criteria are related to the performance of the drainage system, and are based on the drain discharge, the drain capacity, the optimum drain depth, and the spacing, slope and diameter of the drains. The design criteria are [22]:

For the calculation of the depth and spacing of the field drains:
- a design depth of the water table midway between the drains of 1.0 m to guarantee favourable soil-water conditions for the deep-rooting plants (cotton);
- a design discharge of 1.0 mm d^{-1} to maintain the soil salinity below the critical levels for crop production. In the northern part of the Nile Delta, i.e. north of the 3 m$^+$MSL contour, this rate has been increased to 1.2 mm d^{-1} due to upward seepage [149].

For the calculation of the diameters of the field and collector pipe drains:
- a peak design discharge for the determination of drain-pipe capacity of 4 mm d^{-1} for rice areas and 3 mm d^{-1} for non-rice areas;
- a safety factor of 25% in the design of the collector drains to take into account sedimentation and irregularities in alignment;
- no overpressure in the system at discharges equal to the peak design rate;
- a maximum drain depth of 1.5 m for pipe drains and 2.5 m for collector drains.

On basis of these criteria, drain spacings are calculated using Hooghoudt's steady-state approach. In spite of the theoretical computations, a limit is imposed on the drain spacing: minimum 30 m and maximum 60 m [149]. The field drains have an average length of 200 m and a design slope between 0.1 and 0.2%. Collector drains are spaced at 400 m and consist of pipes with increasing diameter. The diameters are based on the Manning equation for transporting pipes using a roughness coefficient derived by Visser [275].

2.5 Installation practices

Organization

Until the end of the 1960s, the Irrigation Department was responsible for the installation of subsurface drainage systems that were constructed, mostly manually and on a limited scale. In the 1970s, Public Excavation Companies (PEC) were established for the mechanical excavation and construction of both canals and drains [150]. These companies belonged to the MWRI, but are now fully owned by the Ministry of Business Development, as a step towards privatisation. They are now part of a separate holding company: *Public Holding Company for Public Works*. Gradually, more contractors from both public and private sectors joined in. The private sector companies started work as sub-contractors (for labour) to public main contractors, and later executed complete projects on their own. To facilitate this, EPADP supplies the contractors, where necessary, with the drainage machinery. Contractors have to pay for the machinery from the instalments due for their work in the projects. When mechanized installation of subsurface drainage systems started some forty years ago, 90% of the contractors were public contractors. Nowadays, the balance has shifted in favour of private contractors.

Drainage materials

Locally manufactured, corrugated PVC pipes with a diameter of 100 mm are used for the field drains [150]. Collector drains are made from concrete or plastic. Traditionally, concrete pipes with diameters between 150 and 600 mm and lengths between 0.75 and 1.00 m were used. The larger diameter pipes (> 400 mm) are reinforced. The introduction of mechanical laying, in the early 1960s for field drains and in the 1970s for collector drains, required different types of pipes. The introduction in 1979 of corrugated PVC pipes significantly helped to boost the progress of Egypt's large-scale drainage projects. A large-scale excavation programme, carried out in the Nile Delta, revealed that sedimentation was significantly reduced after the introduction of plastic pipes [15]. Since 1998, collector drain pipes are made of PVC or HDPE, 200 to 400 mm in diameter and 6 m long. Reinforced concrete pipes are still used, but only at the outlet, the flushing inlet and at places where the collector drains cross roads and irrigation canals.

Concrete collector drains have the same sedimentation problem as the clay pipes. Sedimentation levels in concrete collector drains, with diameters up to 500 mm, reduced the effective cross sectional area by about 35% four years after installation [195]. Thus it is not surprising that plastic collector drains perform better than concrete drains, mainly because of the lower sedimentation rates that offset the higher roughness coefficient caused by the corrugations. The introduction of larger diameter plastic pipes ($150 < \varnothing < 300$ mm) for collector drains took much longer than the introduction of smaller diameter pipes for field drains, mainly because of the complex manufacturing process [152].

The biggest obstacles that had to be overcome for the introduction of corrugated plastic pipes were: (i) the complex manufacturing process, (ii) making the pipes strong enough and flexible, and at the same time keep the weight per metre low, and (iii) the logistic problems, because plastic pipes are more sensitive to temperature and ultra-violet radiation. When exposed to sunlight, the pipes tends to become brittle [3]. Existing standards were updated to include specifications for the new materials from which the pipes are manufactured. These standards, originating from countries with a long drainage history, were adapted to specific, local conditions and circumstances [63].

Natural, graded, gravel is used for envelope if the soil has a clay content of less than 30%. Gravel, however, is costly and difficult to apply. In 1994, pre-wrapped synthetic envelopes were introduced as an alternative for the gravel envelope. Research was conducted in the laboratory, in pilot areas and during normal installation practices to establish the relevant O_{90}-values[4] for the envelopes for the typical range of problems soils that prevail in Egypt (Figure 2.4).

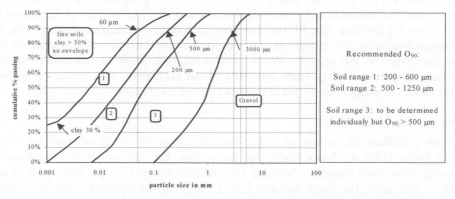

Figure 2.4   Ranges of selected d_{90}-values for use in the Egyptian Nile Delta [63]. Note: the d_{90}-value of a soil based on the sieve mesh (mm) through with 90% of the bare soil material passes [50].

Installation

The implementation of drainage systems involves the following steps [75]:
- construction or remodelling of open main drains;
- construction of drainage pumping stations to keep the water level in the open main drainage system at least 2.5 m below field level so that the piped systems can discharge by gravity;

---

[4] The O_{90}-value is defined as the average diameter of the soil particles in a fraction, 90% of which is retained by an envelope in a standard sieving test based on NNI 1990 [50].

- construction of a composite pipe field drainage systems consisting of field and collector drains.

Since the 1970s, trenchers are used to install field and collector drains with diameters of up to 250 mm. Larger diameter pipes are still installed in trenches dug by excavators. While tractors and trailers transport the pipes and envelopes in the field, manual labour is still used to supply these materials to the machines during installation. The use of manual labour significantly decreased with the introduction of pre-wrapped plastic pipes. The excavation for structures is either done manually or with backhoes depending on the contractor. Backfilling the trenches is mostly done manually although some contractors use tractors equipped with a dozer blade.

Field conditions

In Egypt, the soil type and agro-hydrologic conditions are rather uniform. The majority of the soils consist of relatively deep alluvial soils with high clay and silt content. In three regions, however, different conditions are encountered, i.e. [150]:
- at the fringes of the Nile Valley and Nile Delta, soils tend to contain more sand and lose their structural stability. When the water table is high these soils become problematic particularly when a high hydraulic gradient creates 'quicksand' conditions (Sherashra and Haress areas);
- in the Western Nile Delta, some areas are characterized by calcareous hard rocks in the subsoil (Nubariya area);
- in the northern part of the Nile Delta, the low-lying areas are subject to artesian pressure: significant upward seepage occurs where the resistance of the overlaying low permeable soil decreases with the associated problems of salinity and alkalinity.

For the 'quicksand' areas, special arrangements had to be made to install subsurface drainage. Implementation of the drainage system in the Sherashra area, southwest of Alexandria, was planned to take place in 1974. Auger holes drilled during the field investigation showed a distinct change in the soil profile with unstable light soils below a depth of 1.0 – 1.5 m. As soon as the auger hit the unstable soils groundwater rose under pressure to a shallow depth below the soil surface and the auger holes caved in when digging exceeded the depth of the stable surface soil. Further investigations revealed the prevalence of a piezometric head around soil surface, 1.0 to 1.5 m above the water table (Figure 2.2). A first pilot area implemented at Sherashra produced disastrous results. The concrete pipes used for field drains were soon completely filled with sand. The manually installed collector pipes were dislocated under the effect of quicksand conditions. The area was abandoned and a new pilot area was constructed in Haress, northeast of Sherashra, in 1993-1994 [60]. Pre-wrapped corrugated PVC pipes were used for the field drains and corrugated non-perforated HDPE pipes for the collectors. The field drains were installed successfully and their performance was adequate. However, the results were not entirely satisfactory due to problems with the installation of the gravel envelope. The installation of the collector drains at a greater depth (2.0 - 2.5 m) was again problematic: groundwater rising under pressure in the trench behind the trencher machine made the non-perforated pipe (filled with air) float. The problems were even greater when an attempt was made to lay the bigger pipes in a trench that was excavated with a backhoe. The layers of shells found in the subsoil significantly increased the permeability of the soil at the drain depth.

To overcome these problems perforated pipes were also introduced for the collector drains. During installation these perforated pipes quickly filled with water and consequently stayed in place. A cheap type of envelope (thin sheet) was used to prevent the siltation. Clogging of the envelope was no problem because the collector is not designed to have a dewatering function. The conditions in the Haress area were the reason to test the use of trenchless machines under these conditions [63].

The Nubariya area is part of the Nile Delta's western fringes reclaimed during the 1960s-1970s. The alluvial silty clay topsoil diminishes towards the West and calcareous soil dominates the profile with hard rocks frequently intersecting the soil profile resulting in high water tables. The normal type of trenchers operating in the Delta failed to operate under the Nubariya conditions. A partnership and cooperation between the contractor and the machine supplier yielded a special type of trencher with a more powerful engine and a different type of the digging mechanism [150].

For the low-lying areas in the North, a three-step development approach was developed [53]:
- during the first 1 to 3 years after reclamation, surface drainage is installed and halophytes (salt-tolerate crops) are cultivated. Gypsum or other amendments are applied to improve of the top 10 - 20 cm of the soil profile;
- after 3 to 5 years, mole drains are installed and salt resistant/tolerant crops are cultivated to improve soil structure and fertility by nitrogen fixation. If required, more gypsum is applied;
- finally, after the heavy clay soils have ripened and reached a hydraulic conductivity greater than 0.1 m/day, subsurface drains can be installed at economical spacing. Subsurface drainage, in combination with the existing surface drainage, enables the cultivation of more profitable, i.e. less salt-tolerant, crops.

In these successive stages a close cooperation between technical (infrastructure, drainage and soil improvement), agronomic and social disciplines is needed. Farmers are responsible for the construction of the surface drainage and the management of the field system. To achieve this, appropriate technologies are made available to the farmers.

2.6 Disposal of the drainage effluent

The River Nile is not only the source of irrigation water in Egypt, it also is the main disposal drain as all drainage effluent from the agricultural lands in the Nile Valley is discharged back to the river. This is possible because only about one third of the agricultural lands are located in the Nile Valley (Figure 2.1). Of the total amount of water passing the Aswan High Dam (approximately 55×10^9 m^3y^{-1}) about 20×10^9 m^3 y^{-1} is used to irrigate the agricultural lands between Aswan and Cairo (approximately 0.9 Mha). Because all the drainage water is discharged back into the River Nile, the salinity of the Nile water increases in downstream direction (Table 2.1). This practice is safe and sustainable because the salinity of the water entering the Nile Delta is still so low (< 0.47 dS m^{-1}) that it can be used for irrigation. In the Nile Delta, however, a separate main drainage system had to be constructed to discharge the drainage effluent directly in the sea because diverting this water back to the river would result in unacceptable high salinity levels.

Table 2.1   Discharge, salinity, and salt load in the River Nile [196]

Location	Discharge $(10^9\,m^3\,y^{-1})$	Salinity $(dS\,m^{-1})$	Total salt load $(10^9\,kg\,y^{-1})$
Aswan High Dam	55	0.31	11.0
Delta Barrage Cairo	35	0.47	10.5
Mediterranean Sea	14	3.59	32.0[a]

[a] The increase in the total salt load between Cairo and the Mediterranean Sea is due to the leaching of deeper (saline) soil layers and the seepage of saline groundwater.

Since 1930, 21 pumping stations have been built in the Nile Delta to pump part of the drainage water back into the irrigation system. In the mid 1970s drainage water reuse became an official policy and a component of the national water resources plan [7]. In 1996/97, 4,400 million m³ of drainage water with an average salinity of 1.8 dS m⁻¹ was pumped back into the irrigation system [121]. At field level, farmers also reuse drainage water for irrigation by pumping it directly from the drains. This 'unofficial' reuse is estimated to be between 2,800 and 4,400 million m³ per year. Both the official and unofficial reuse cover about 15% of the crop water requirements [72; 196]. The total estimated reuse potential is 9,700 million m³ with a maximum salinity of 3.5 dS m⁻¹, of which 8,000 m³ can be used effectively [121].

A major disadvantage of reuse is that, because the salinity of the reused water is often high, it contributes more than proportionally to the total salt supply to the crop. It is estimated that the contribution of the 15% reused water is about 46% of the total salts supplied through irrigation [16]. A monitoring programme conducted by DRI showed that areas where drainage water is reused have slightly higher soil salinity levels compared to areas that are irrigated by fresh water only [61]. The soil salinity levels, however, remained within the tolerable range for most crops, but yields were 9 to 15% lower [7]. Another problem is that the pollution of the drainage water has increased since the 1990s. The water is not only polluted by remains of dissolved nitrates and/or fertilizers leached out through the subsurface drainage system but also by untreated municipal and industrial waste water.

In Mashtul pilot area, the leaching of agro-chemicals form fields cultivated with various crops and drain intensities was monitored [10]. The results show that the concentration of nitrates fluctuates during the seasons with a remarkable increase after each fertiliser application. Pollution of the shallow groundwater with nitrates (NO_3) during both the winter and summer season is very similar to the pollution of the drainage water (discharge from the field drains). The concentration of nitrates in the drainage water, however, is very much influenced by the drainage intensity. The concentration in drainage water from fields with deeper drains (1.50 m) and narrow spacing (15 m) reached higher peaks than those of the shallower and wider drain depth/spacing combinations. The nitrate concentration in the groundwater and drainage water during winter is small and seldom exceeded 25 ppm, because *berseem* is not fertilised with nitrates. Consequently, the nitrate concentration in the collector and open main drains was small. The nitrate concentration in the drainage water of rice fields is relatively less compared to the other summer crops. Continuous flooded crops produce lower concentrations than intermittently irrigated crops probably due to dilution and denitrification. The concentration of nitrates in the drainage water is also reduced as the drainage water flows from the field drains into the collectors and then to the main drain. In the collector system, the field drainage water from different field crops gets mixed together. The open main drain usually receives direct irrigation water losses and surface runoff which causes further dilution of the nitrates concentration. The peak nitrates concentration during summer at the outlet of the collector and the open main drain were 152 and 89 ppm, respectively. The concentrations, however, are still so high that aquatic weed growth is enhanced.

2.7   A modified layout of the subsurface drainage system for rice areas[5]

Introduction

In Egypt, crop rotation is practised with wheat and *berseem* in winter, and cotton, maize and rice in summer, in addition to vegetables, orchards, and sugarcane (Figure 2.5). The conventional layout of the subsurface drainage system used in Egypt is of a composite type, consisting of lateral and collector drains, not adapted to the crops grown in the fields (Figure 2.6). The drainage criteria are based on the most critical crop grown in the area (cotton). The design rate for collector drain pipe capacity is 3 mm d^{-1}, for rice growing areas this rate was increased to 4 mm d^{-1} to enable adequate drainage conditions for the 'dry-foot' crops [22].

In spite of this increase in the design rate, water management problems occur in areas where rice is cultivated along with 'dry-foot' crops [12]. Rice is the only crop with water standing on the soil surface, and consequently rice fields suffer huge water losses through the subsurface drainage system. To save irrigation water, farmers block the collector drains at the nearest manhole with whatever is available, i.e. straw, mud, etc. As a result, the 'dry-foot' crops in the upstream part of the drainage area may suffer from waterlogging.

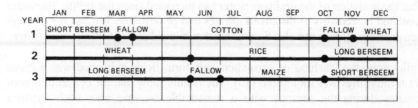

Figure 2.5   Crop rotation practices in Egypt, example from Mashtul pilot area

---

[5] Published as: El-Atfy, H.E., Abdel-Alim, M.Q., Ritzema, H.P., 1991. A modified layout of the subsurface drainage system for rice areas in the Nile Delta, Egypt. *Agricultural Water Management*, **19**, 289-302

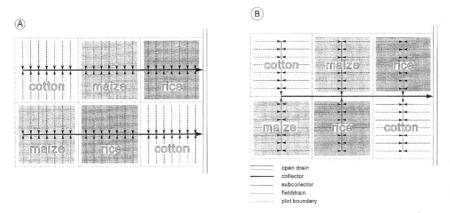

Figure 2.6   Conventional (A) and modified (B) layout of the subsurface drainage system used in Egypt

After the rice seasons, the man-made 'plugs' often remain in the collector drain, causing maintenance problems. To translate these farmers' practices into a technically sound and environmentally safe subsurface drainage system the concept of a modified layout has been developed [71]. The main features of the concept are:

- to restrict the outflow from the areas cultivated with rice;
- to enable normal drainage conditions for the remaining areas (cultivated with '*dry-foot*' crops).

The modified layout is based on the crop consolidation system, which has been practised in Egypt since 1960. According to this system, crops are grown in units with fixed boundaries. The modified layout consists of a main collector drain with several subcollector branches (Figure 2.6). The design criteria within a subcollector area (e.g. depth and spacing of the lateral drains) remained unchanged as they are still based on the growing conditions of the most critical 'dry-foot' crop (cotton). Each subcollector coincides with one crop consolidation unit and is equipped, at the junction with the main collector, with a closing device to regulate the subcollector outflow. If rice is cultivated in the drainage area of a subcollector, the outflow of drainage water is restricted by closing this device. If any other crop is grown, the subcollector is left open, enabling 'dry-foot' crop drainage conditions. As a consequence, the design rate for collector drain pipe capacity could be reduced to 3 mm d^{-1}, the design rate for non-rice areas.

The new concept was introduced on a pilot scale in Mahmoudiya in the eastern Nile Delta in 1982. The farmers showed great interest in having closing devices to control the outflow from their rice plots. Interviews and discussions with the farmers revealed their acceptance of, and preference for, the modified system [166; 167]. However, it was necessary to be sure that the introduction of the modified layout would have no negative effect on soil salinity and crop yield, to ensure the availability of the basic data required for the design (crop consolidation maps), to clarify the operation requirements, and to evaluate the cost. Therefore a monitoring programme to study the effect of the introduction of the modified subsurface drainage system started in 1983. In this section the results of this monitoring programme, which continued for six years, are presented.

Monitoring programme and methods

The principles of the modified layout were tested at several locations, each representing a major rice-growing area in the Nile Delta (Figure 2.7). The objectives of the monitoring programme were to obtain a better insight into [67]:

- the reliability of the crop consolidation maps;
- the effects of the water management practices on crop production and soil salinity;
- the effects of the water management practices on the performance of the subsurface drainage system;
- the operation and performance of the closing devices in the modified drainage system.

The investigations were conducted at three levels: (i) fully controlled experiments at three experimental field stations, (ii) in-depth studies in farmers' controlled fields, and (iii) large-scale monitoring programmes in the two pilot schemes. At each location the soil and hydrological characteristics as well as the farmers' practices were assumed to be identical for the adjacent modified and conventional areas, the only difference being the restricted outflow of the subcollectors of the units in the modified system cultivated with rice. The controlled experiments were conducted in Zankalon (East Delta), Sakha (Middle Delta) and King Osman (West Delta).

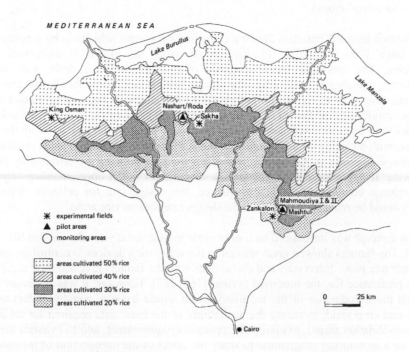

Figure 2.7   Location of the study areas

The studies in the farmers' fields were conducted in Mashtul (110 ha), which is part of the first pilot scheme Mahmoudiya, an area of around 1,600 ha in the eastern Nile Delta, and in Nashart/Roda, which is part of the second pilot scheme Nashart, an area of 2,100 ha in the middle of the Nile Delta [14]. The monitoring programme covered the following parameters:

- crop pattern and crop intensities;
- irrigation water applications (only in the experimental fields) and the daily drop of the standing water in the rice fields;
- collector discharges and salinities;
- performance of the closing devices;
- soil salinity level before and after the growing season;
- crop yield of rice, maize and cotton;
- cost of the modified system in comparison with the conventional layout.

Results and discussions

Crop pattern and crop intensities

The crop consolidation is the backbone of the modified system. The actual cropping patterns in two project schemes were compared with the crop consolidation maps on which the design of the subsurface drainage system is based. In Mashtul pilot area, which is part of the first pilot scheme (Mahmoudiya), the actual cropping pattern was surveyed from 1981 through 1984. The area cropped in accordance with the crop consolidation map was found to range between 96 and 100% [167]. For the total Mahmoudiya area, the actual cropping pattern in the summer of 1985 was compared with the crop consolidation map. Discrepancies occurred in 9% of the total area, although only 3% caused operational problems, namely when rice was cultivated with either maize or cotton in the same subcollector. The remaining 6% was a mixture of cotton and maize, both in need of unrestricted drainage outflow. Discrepancies were mainly observed in small areas surrounding villages. It should be emphasized that a change of crop within a complete crop unit does not violate the operation of the modified drainage system. On the basis of these findings, the subcollectors in the second pilot scheme (Nashart) were designed in cooperation with the Agricultural Department in Kafr El Sheikh [168]. Slight modifications of the crop consolidation units reduced the number of subcollectors needed. The results of the monitoring programme of 1988 showed that also in this area only a negligible percentage (3%) of the area was cultivated with rice along with either cotton or maize [174]. It can be concluded that the crop consolidation maps are a sound and reliable basis for the design of the modified layout. The required information is easily obtainable at the agricultural departments at district level.

Irrigation water applications

Water management practices in rice fields differ greatly between the modified and the conventional systems. Rice fields under modified drainage conditions require less irrigation water to maintain the same height of ponding water because of the restricted outflow of the subsurface drainage water. During the summer season of 1984 fully controlled water management experiments were conducted in modified and conventional units at three experimental fields [170]. All units were cultivated with rice under optimum water management conditions; if the average height of the standing water dropped below 5 cm, irrigation water was supplied to a level of 9 cm. The irrigation water supply as well as the daily drop in standing water were measured in both systems. The total irrigation applications over the cropping season are presented in Table 2.2.

Table 2.2   Total irrigation water applications to rice fields under optimum water management conditions in the three experimental fields

Subsurface drainage system	Irrigation application (m^3 season^{-1} ha^{-1})		
	Zankalon	Sakha	King Osman
Conventional	16,000	13,800	22,700
Modified	8,800	7,200	15,500

The differences in water use between the three experimental fields are due to the different hydrological conditions. Nevertheless, in all three areas, the fields with a modified layout required around 40% less irrigation water than the conventional units. To check if a relation exists between the irrigation application (which could be measured in the experimental fields only) and the daily drop of the standing water (which could be measured in both the experimental fields as well as the farmers' field) the seasonal average values were calculated. Figure 2.8 shows that a good agreement exists in the experimental fields between the measured irrigation applications and the daily drop in standing water. Consequently, it can be concluded that in farmers' fields, where it was not possible to measure irrigation water applications because farmers used portable engine driven pumps, the rate of the daily drop in the standing water is a good indication of the irrigation water applications.

Farmers did not manage to maintain optimum water conditions because of irrigation water shortages; this occasionally resulted in no standing water at all, especially in fields with a conventional layout. As the daily drop was averaged over the whole cropping season it seems that the drop is much lower in the farmers' fields compared to the experimental fields. If the data had been corrected for the number of dry days the differences would have been less. In the farmers' fields, however, the daily drop in the modified units was again less than the daily drop in the conventional units, the difference being between 22 and 35%. It can be concluded that, under normal farming practices, the average saving in irrigation water supply to the rice plots in the modified system is around 30%. As a consequence, farmers in the modified system need a less frequent rate of irrigation water application, which also implies savings in operational activities [169].

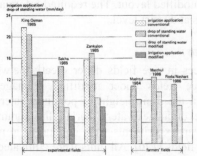

Figure 2.8   Irrigation application and daily drop of the standing water.

Collector discharges and salinities

Discharges were measured at the outlet of the collector drains with a bucket and stopwatch and simultaneously the salinity of the drainage water with a portable conductivity meter. Each year the measurements started in June and continued until October, covering the total summer season. Each collector was measured at least once a week [174]. Table 2.3 presents the seasonal average discharge rates, together with the corresponding salinities. The average salt removal through the collector system was calculated from the mean discharge rates and mean salinity levels.

The data from the different areas are in good agreement with each other: differences in water management practices clearly result in lower discharge rates in the modified systems. The analysis of the individual collector drains [167; 172; 175] showed that in the modified systems the 90% cumulative discharge rate was independent of the rice intensity in the drainage area. This is in contrast to the conventional units, where the average 90% cumulative discharge rate increased from approximately 2.0 mm d^{-1} for collectors with less than 20% rice to approximately 3.5 mm d^{-1} for collectors with more than 60% rice. The discharges from collectors with less than 20% rice were in the same order of magnitude as the discharges from the modified collectors, although in the latter the rice intensity was much higher. It is clear that the introduction of the modified system reduced the discharge through the collector drains. As a consequence, the design rate for collector drain capacity (3 mm d^{-1}) can be reduced even further, as the discharge rate at 90% cumulative frequency did not exceed 2.3 mm d^{-1} (Table 2.3).

Performance of closing devices

The function of a closing device is to restrict the outflow of a subcollector serving an area cultivated with rice. Four types of closing devices were tested, namely the steel flap gate, the steel sliding gate, the aluminium disc plug, and the wooden plug [166; 172]. The steel flap gate emerged as the most promising device and was tested on a large scale.

Table 2.3   Average seasonal collector discharge rates and salinities

Area	Summer season	A_r[a] (%)	Q_{90}[b] (mm d^{-1})	Q_{mean} (mm d^{-1})	EC_{mean} (dS m^{-1})	TS[c] (kg ha^{-1} d^{-1})
Modified system:						
Mahmoudiya I	1984[d]	41	1.4	1.0	2.3	14.8
Mahmoudiya I	1987	17	1.0	0.7	2.4	10.7
Nashart	1986	20	2.3	0.9	2.9	16.7
Nashart	1988	55	2.1	1.1	3.1	21.9
Conventional system:						
Mahmoudiya II	1984	15	1.7	1.0	2.2	14.0
Roda	1986	35	3.0	1.8	3.4	39.3
Roda	1987	36	2.7	2.0	3.1	39.8
Roda	1988	53	2.8	2.1	3.2	43.1

[a] A_r = part of the drainage area cultivated with rice
[b] Q_{90} = discharge rate at 90% cumulative frequency
[c]   TS = average total salt removal over the summer season
[d]   25% of the area had no restricted outflow conditions

A total of 31 flap gates were installed in the Mahmoudiya pilot area during the spring of 1985 and their performance was monitored during the following summer season [168]. A total of 31 flap gates were installed in the Mahmoudiya pilot area during the spring of 1985 and their performance was monitored during the following summer season [168]. The performance was evaluated by the difference in water level between the upstream and downstream manholes and by regular visual inspection. The performance of 73% of the gates was rated as good, bad performance being mainly due to difficult installation conditions (submerged outlets of the collector pipes).

In the spring of 1988, the same prototypes were installed in the other project area (Nashart). The best performance observed was again by the steel flap gates and, to a lesser degree, by the wooden plugs [174]. Although some leakage occurred, the outflow from subcollector areas cultivated with rice was considerably reduced. Neither the sliding gates (too much leakage) nor the aluminium disc plugs (pushed out by the water pressure) performed satisfactorily. From an operational point of view, the steel flap gates are preferred above the wooden plug, because they are permanently installed, whereas the wooden plugs have to be removed after each rice season. To guarantee a satisfactory performance, it is important to install the gates during construction of the subcollector drains, or at least under dry conditions. Operation of the gates does not require special skills and can be done by either the farmers or the maintenance teams.

Soil salinity

Soil samples were collected two times during each summer season [174]. The first sampling was done just before the start of the rice season and the second sampling during harvest. The soil samples were collected from three layers, i.e. the surface layer (0 - 25 cm), the subsoil surface (25 - 50 cm), and at drain depth (125 - 150 cm). The salinity of the saturation extract was measured at the soil laboratory of DRI. To compare the data, the average salinity level over the top 0.50 m of the soil was calculated, the results are presented in Table 2.4. In spite of the reduced irrigation applications and the corresponding lower drainage rates, sufficient leaching took place in the modified units to keep the soil salinity levels well below the critical value of rice, i.e. $EC_e < 3$ dS m^{-1} [59]. This critical value is defined as the average level of the soil salinity above which reduction in yield will occur. For the prevailing soil and hydrological conditions in the Nile Delta and the rice varieties used in Egypt the critical value is even slightly higher, around 3.5 dS m^{-1} [8].

Regardless of the type of subsurface drainage system (modified or conventional), the level of the standing water in the rice fields results in a downward flux of the water in the soil, which is sufficient to maintain favourable soil salinity levels. Furthermore, in the modified system, the closing devices are opened at the end of the rice season to drain off the excess water. In the recently constructed projects, the decrease in soil salinity levels was the same for the conventional and modified units, although the total amounts of salts removed by the subsurface drainage system were much higher in the conventional units (Table 2.3). This can be attributed to the higher irrigation water requirements in the conventional units and the fact that the salinity of this irrigation water is relatively high due to the occasional reuse of drainage water in periods with water shortages.

Table 2.4 Average soil salinity levels before and after the rice growing season

Area	Year	Average soil salinity (dS m^{-1})[a]	
		Before	After
Modified system:			
Mahmoudiya I	1983	2.07	1.98
Mahmoudiya I	1984	1.26	0.85
Mahmoudiya I	1985	1.05	1.40
Zankalon	1985	1.00	0.69
Mashtul	1986	1.08	1.32
Sakha	1985	3.26	1.92
Nashart	1986	2.79	1.41
Nashart	1988	1.84	1.43
King Osman	1985	2.75	1.34
Average modified systems		1.90	1.37
Conventional system:			
Mahmoudiya II	1983	1.08	3.02
Mahmoudiya II	1984	1.27	1.03
Mahmoudiya II	1985	1.35	1.35
Zankalon	1985	1.34	0.80
Mashtul	1986	1.20	1.22
Sakha	1985	3.29	1.42
Roda	1986	3.25	1.51
Roda	1988	1.39	1.30
King Osman	1985	2.48	1.21
Average conventional systems		1.85	1.43

[a] average over the top 0.50 m of the soil profile

Crop yield

The yield of any crop is a function of a combination of factors (e.g. agricultural inputs and practices, soil and hydrological conditions), which makes it difficult to quantify the influence of each parameter separately. Nevertheless, it is assumed that in each selected area these factors are the same for the modified and conventional units, the only difference being the type of subsurface drainage (modified or conventional). So it is possible to compare the yield figures obtained in the fields with a modified system with those from the fields with a conventional system. Samples of the rice and maize crops were taken from the same locations used for the soil samples. Data on the yield of cotton were obtained from the Agricultural Cooperatives. The yield of the individual fields was characterized by high variability, which is not surprising because of the many factors that influence crop production. For both the conventional and the modified units of each area, the average yield figures per area per season were calculated [174]. They are presented in Table 2.5 and the overall averages in Figure 2.9. Although there is some variation between the seasons and between the areas, no significant differences could be established between the modified and the conventional units.

Table 2.5   Crop yield in the modified (mod) and conventional (con) systems

Area	Year	Average yield (t ha^{-1})					
		Rice		Maize		Cotton	
		mod.	con.	mod.	con.	mod.	con.
Mahmoudiya	1983	6.2	5.0	4.5	3.3	3.1	2.6
Mahmoudiya	1984	6.2	6.0	6.7	3.8	3.1	2.9
Mahmoudiya	1985	5.5	7.1	5.2	5.0	3.6	3.3
Mashtul	1986	5.0	5.0	-	-	-	-
Zankalon	1985	4.3	5.2	-	-	-	-
King Osman	1985	5.2	5.0	-	-	-	-
Sakha	1985	8.1	7.1	-	-	-	-
Roda/Nashart	1986	3.8	6.0	-	-	-	-
Roda/Nashart	1988	6.7	6.4	4.5	5.5	2.4	2.6
Overall average		5.7	5.9	5.2	4.4	3.0	2.9

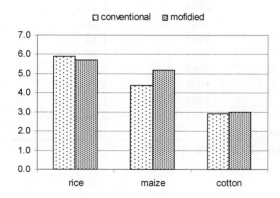

Figure 2.9   Overall average crop yield.

Cost comparison between modified and conventional systems

For the two large pilot projects (Mahmoudiya I and Nashart) both, a conventional and modified design were made. On the basis of unit prices, the differences in construction costs between the two systems were calculated [166; 168]. The total length of pipes in the modified system is greater because of the introduction of subcollectors which, together with the installation of closing devices, leads to extra costs. On the other hand, the lower design rate implies a reduction in the size of the collector pipes as compared to the current design norms and thus leads to cost savings. Savings in maintenance costs and the benefits of a more reliable system have not been considered in the analysis. Mahmoudiya I was the first area where the modified system was introduced on a large scale; it was constructed in 1982. Based on 1983 prices, the costs of the modified system were 12% higher than those of the conventional system. This difference can be attributed to:

- the relatively small size of the subcollector units in the modified system;
- the design rate for the collector drains in the modified system was 3 mm d^{-1}, which is quite high compared with the design rate for non-rice areas (2 mm d^{-1}). The design rate for the conventional system was 4 mm d^{-1}.

Based on the experiences obtained in the Mahmoudiya area, the design of the Nashart area was slightly adjusted. In cooperation with the Agricultural Department in Kafr El Sheikh, minor modifications were made in the crop consolidation scheme to reduce the number of small subcollectors. Also the design rate of the collector drains in the modified system was reduced to 2 mm d^{-1}. Based on the prices of 1985, the costs of the modified system at Nashart were approximately 6% lower than those of the conventional system.

Conclusions

To reduce irrigation water losses from rice areas without restricting the subsurface drainage from 'dry-foot' crops, a modified design for the subsurface drainage system is required. The modified system, which is based on the crop consolidation scheme, was tested at several locations in the Nile Delta. After the principles were studied in experimental fields, detailed investigations were carried out in farmers' fields and followed-up by large-scale monitoring programmes. The study covered a six-year period, running from 1983 to 1988. The introduction of the modified layout of the subsurface drainage system in rice-growing areas in the Nile Delta resulted in:

- savings in irrigation water up to 30%. This irrigation water would otherwise be discharged through the subsurface drainage system: the difference in drainage rates from rice fields between the conventional and modified drainage system amounts of 1 to 3 mm d^{-1} over a growing season of approximately 100 days;
- protection of the drainage system from justifiable, although unauthorized and improper, interference by farmers to stop irrigation water losses from rice fields through the subsurface drainage system, and thus reduce the maintenance requirements;
- protection of crops other than rice from the damaging effects of improperly blocked conventional collector drains.

These benefits were obtained without any negative effects on either soil salinity or crop yield and with no increase in costs compared with the conventional system. Despite these positive effects, EPADP did not install the modified system to most rice areas in the Delta, mainly because the consolidated cropping pattern was abandoned in the 1990s. The modified concept, however, can enhance the participatory approach in water management recently adopted by MWRI.

2.8   Controlled drainage and farmers participation

To increase awareness about the benefits of the modified subsurface drainage system and to promote acceptance of this technique by farmers and the authorities, the modified drainage was introduced as controlled drainage in 1995 [17; 63]. Through traditional field trials, using participatory rural appraisal techniques and advertising the opportunities. Emphasis was put on: (i) farmers involvement in operation and (ii) savings in irrigation water supply. Farmers were organized on voluntary basis to consolidate the rice areas and the collectors were provided with closing devices. Observations were also made along two other collector drains where farmers did not consolidate rice areas. The results show that the average irrigation water supply for the modified system is 1,805 m^3 ha^{-1} compared to 3,169 m^3 ha^{-1} for conventional collectors. This means that the modified drainage system saves about 43% of irrigation water compared to the conventional system, reducing the costs of renting pumps with a similar percentage. The reduction in irrigation water supply did not result in a yield decrease (the average yield in the areas with a modified system was 2.7 t ha^{-1} compared to 2.6 t ha^{-1} in the areas with a conventional system). The decrease in soil salinity indicated that although the subsurface drainage in the areas with a modified system is restricted, the leaching requirements are still met. Based on the findings, guidelines to help with appropriate design and water management have been prepared [62].

2.9   Verification of drainage design criteria in the Nile Delta[6]

Introduction

Since the beginning of the 20th century the use of water per unit area in the Nile Delta of Egypt has increased sharply with the gradual introduction of perennial irrigation. Consequently the natural drainage system could no longer cope with the increased percolation losses and land became waterlogged and/or salt-affected. To overcome these problems the Egyptian Government is implementing an intensive land drainage programme to provide the whole Nile Delta with subsurface drains. The original design criteria were established in the early sixties [4]. To verify these criteria for the south-eastern part of the Nile Delta a drainage pilot area was constructed at Mashtul in 1979-1980 [171]. Mashtul is situated 70 km north-east of Cairo in a rather flat area, characterized by a, relatively homogeneous, clay layer on top of a sandy aquifer. The clay layer, which is approximately 6 m thick, contains about 35% silt and 65% clay. The area represents the prevailing soil, hydrological, and agricultural conditions of the south-eastern Nile Delta. The climate is characterized by a long dry summer and a short winter with little rainfall (annual average in the range of 50 to 100 mm). The reference crop evapotranspiration, calculated with the modified Penman method, ranges from approximately 3.5 mm d^{-1} during winter to 6.0 mm d^{-1} during summer [173]. A 3-year crop rotation, including wheat and *berseem* in winter, and rice, maize and cotton in summer, is practised in the area (Figure 2.5). The units are cultivated with a single crop during each cropping season.

The area is irrigated on rotation by surface flooding; the average irrigation interval varies between 23 and 28 days for the winter crops and between 5 and 22 days for the summer crops. In the latter case, rice is irrigated at short intervals while cotton irrigation intervals are the longest. The subsurface drainage system consists of parallel PVC pipe drains, which discharge into buried concrete collector drains. The area is divided in 18 drainage units with different drain depths and spacings (Figure 2.10). The pipe drains have an average length of 200 m and a design slope of between 0.1 and 0.2%. Collector drains are spaced at 400 m and consist of pipes with increasing diameter.

Monitoring programme

The monitoring programme started in 1977, three years prior to the installation of the subsurface drainage system, and continued up to the winter season 1986/1987 [171]. The

---

[6] Published as: Abdel-Dayem, S., Ritzema, H.P., 1990. Verification of drainage design criteria in the Nile Delta, Egypt. *Irrigation and Drainage Systems*, **4**, 117-131.

programme included the monitoring of the cropping pattern, crop yield, soil salinity, depth of the water table, discharge and salinity of the pipe and collector drains, and the water levels in the manholes of the collector system. The programme was conducted in farmers' fields and did not interfere with their agricultural and water-management practices. The cropping pattern during each season was monitored, and crop and soil samples were collected during harvest and analysed according to the standard procedures of the DRI - the only exception being data on cotton yields, which were collected from the local farmers' cooperatives.

The soil samples were taken from the layers 0 - 25 cm and 25 - 50 cm. Starting with the winter season of 1982/1983, the depth of the water table was measured daily in eight selected drainage units (Figure 2.10). Observation wells (perforated PVC pipes, 25 mm diameter) were installed midway between the pipe drains in three rows of three wells, each row at a distance of respectively 50, 100 and 150 m from the outlet of the pipe drains. The wells were installed up to a depth of 0.3 m below drain level, and water table levels were measured daily with a sounder. In the same units, the pipe discharges were measured daily with a bucket and stopwatch. Each collector drain was equipped at its outlet with a V-notch and an automatic water-level recorder to monitor the discharges. Salinities of the discharges were measured daily with portable electrical conductivity meters. Water-level recorders were installed in four selected manholes to monitor the occurrence of overpressure in the collector drains. The analysis of the different parameters and their relationships are based on daily averages of the measured parameters. Subsequently daily values were used to calculate averages per unit and per season. As the start and end of the growing season depends on the crop (Figure 2.5), the winter season was considered to run from November to May and the summer season from June to October.

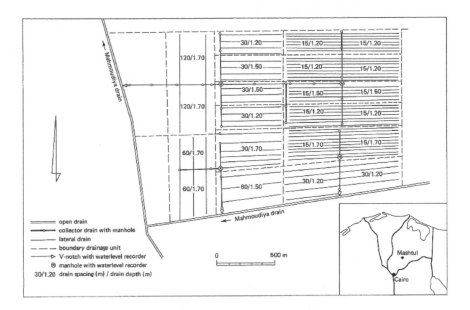

Figure 2.10   Layout of the subsurface drainage system in Mashtul pilot area

Results and discussion

Crop production

The average yield before and after installation of the subsurface drainage system was determined for each of the main crops cultivated in the pilot area [173]. These are plotted in Figure 2.11. The increase in yield was 10% for rice, 48% for *berseem*, 75% for maize, and more than 130% for wheat. Unfortunately, no control area was incorporated into the design of the area, so it is difficult to prove that the increase is solely the result of the improved drainage conditions. Besides, it is clearly understood that the measured increases are the result of several improved inputs, one of which is drainage [13].

Depth of the water table

Only limited data on the water table are available from the pre-drainage period. Single measurements were taken in a network of auger holes over the entire area. The water table was relatively deep, at an average depth of 1.33 m [171]. The available data are not enough to establish a relation between crop yield and the depth of the water table during that period. After the installation of the subsurface drainage system, the water table showed great fluctuations with time [173]. It reached its highest level shortly after irrigation, when it often rose to the soil surface, and then started to fall at a much slower rate (Figure 2.12).

For each growing season and for each drainage unit, the average depth of the water table was plotted versus the average yield (Figure 2.13). Rice was excluded from this analysis, because the water table was not monitored during the rice season. Figure 2.13 shows that the yields of wheat, cotton and maize were not affected by the seasonal average depth of the water table. Only *berseem* suffered a decrease in yield when the average water table dropped to a depth of 0.90 m. It is possible that this yield reduction is related to lack of soil moisture under the existing irrigation and drainage conditions. Nijland and El-Guindi [151] found that for similar areas in the Nile Delta the yields of cotton and wheat were not influenced by seasonal average water table depths of, respectively, more than 0.90 and 0.40 m. It can be concluded that for areas and cropping patterns like Mashtul, the optimum seasonal average depth of the water table is around 0.80 m. At this depth the yield of cotton, maize and wheat is not affected and if the water table is deeper, the yield of *berseem* may suffer from deficiency in soil moisture. Therefore, the design depth of the water table midway between the drains can be reduced from 1.0 to 0.80 m without fear of yield reduction.

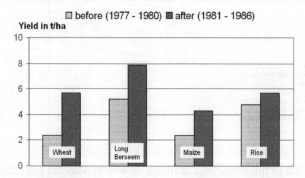

Figure 2.11 Average yield before and after the installation of subsurface drainage.

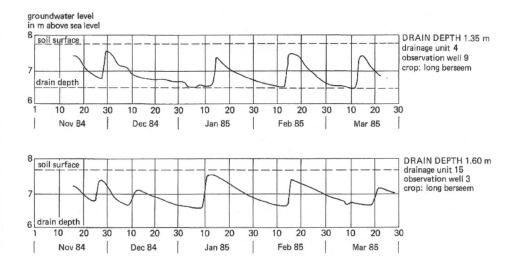

Figure 2.12   Two typical examples of hydrographs showing the fluctuation of the water table.

Drain depth

Usually, deeper drains should allow wider drain spacings because of the increased hydraulic head midway between drains, provided that the outlet is deep enough. Similarly, if the spacing is kept constant but the drains are installed at a greater depth, the average water table should be lower. Table 2.6 presents the seasonal average depth of the water table for each of the six combinations of drain spacings and drain depths in the pilot area. As can be seen from this table, deeper drains do not necessarily result in a lower average water table, especially for the 30 m spacing and drain depths below 1.40 m. It can be concluded, that for similar soils as those found in Mashtul, drains at a depth of 1.20 - 1.40 m are an optimum choice.

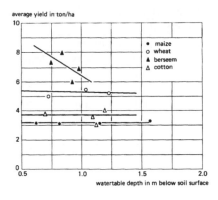

Figure 2.13   Relation between the average depth of the water table and the crop yield per drainage unit and per growing season

Table 2.6   Seasonal average depth of the water table for the different combinations of crops, drain spacing and drain depths.

Spacing (m)	Depth of the water table (m)					
	15			30		
Drain depth (m)	1.20	1.40	1.60	1.20	1.40	1.60
Wheat	0.91	1.15	1.11	0.78	0.95	0.96
Berseem	0.84	1.00	0.98	0.81	0.81	0.75
Maize	0.68	0.87	1.01	0.74	0.82	0.69
Cotton	-	1.13	1.20	0.70	-	1.10

Soil salinity and crop yield

The effect of soil salinity on yield greatly depends on the type of crop. Before the installation of the subsurface drainage system, the soil salinity showed a wide variation (Table 2.7). For each crop, the relationship between yield and average soil salinity was analysed with the method described by Nijland and El Guindi [151]. An example is given in Figure 2.14, which shows the relation between the yield of wheat and the average soil salinity before (Figure 2.14a) and after (Figure 2.14b) the installation of the subsurface drainage system. The upper envelope in the figure represents the maximum yield and the lower envelope the minimum yield. Before the installation of the subsurface drainage system, a decrease in yield occurred when the soil salinity was above a critical value (breakpoint). This breakpoint is not an absolute value, but varies from year to year and from field to field, because the yield also depends on other agricultural inputs. For each crop, the average value of the breakpoint was calculated by combining the data for all seasons. The results found in Mashtul pilot area were compared with data from other sources, as listed in Table 2.7. There are some differences between the breakpoints as found by FAO (world-wide averages) [31], for the Nile Delta as found by Nijland and El Guindi [151] and the values at Mashtul. The differences can be attributed to factors such as farm inputs, soil and hydrological conditions, etc., each influencing the crop yield. After the construction of the subsurface drainage system, the soil salinity decreased. To estimate the effect of the desalinization on the crop production, a two-step approach was followed.

Table 2.7   Soil salinity and the critical values of the crops grown in Mashtul pilot area

Crop	Soil salinity (dS m^{-1})				Critical value*		
	Before drainage		After drainage				
	Mean	Max	Mean	Max	Mashtul	Delta	FAO
Wheat	4.6	11.0	2.1	4.1	3.2	5.5	5.7
Berseem	3.6	9.5	1.4	4.6	3.1	2.5	1.5
Maize	2.7	6.2	1.1	4.0	2.4	3.0	1.7
Rice	2.1	7.8	1.1	3.8	-	3.5	3.0
Cotton	-	-	-	-	-	> 7.0	7.7

* maximum soil salinity level without yield reduction, see references in the text

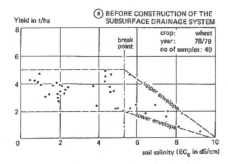
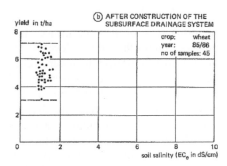

Figure 2.14   Typical example of the relation between crop yield and soil salinity before (a) and after (b) the installation of the subsurface drainage system

Step 1: the data on crop yield and soil salinity over the period before the installation of the subsurface drainage system were used to calculate:
- the average crop yield for the fields with soil salinity levels below the breakpoint;
- the overall yield over this period, thus including the fields with high soil salinity levels;
- the difference between these two values represents the expected increase in crop yield due to the desalinization induced by drainage and under the assumption that all other agricultural inputs remained unchanged.

Step 2: the actual increase was calculated by comparing the yield figures of the period before (1976 - 1980) and after (1981 - 1987) the installation of the subsurface drainage system.

Table 2.8 presents the expected increases in crop yield of wheat, *berseem*, maize and rice due to desalinization effects of drainage and the actual increase, as occurred after the installation of the subsurface drainage system. That the expected increase is less than the actual increase can be attributed to the combined effect of the improved drainage conditions (of which desalinization is only one) and several improved agricultural inputs. It was only after the installation of the subsurface drainage system that many innovative agricultural inputs (like new seed varieties, fertilizers, and other farming practices) became effective.

Table 2.8   Expected (due to desalinization) and actual (due to the installation of the subsurface drainage system) increases in crop yield.

Crop	Increase in yield (%)	
	Expected	Actual
Wheat	56	138
Berseem	37	48
Maize	39	75
Rice	-	10

Average drain discharges

The discharges of the subsurface drainage system, in combination with the natural drainage, control the salinity level of the soil. Piezometric observations revealed that the Mashtul area is subject to natural drainage [173]. The analysis of the water and salt balances over the period November 1983 to October 1986 indicated that the natural drainage rate may reach 0.5 mm d^{-1} [2]. Soil salinity levels after the installation of the drainage system show that the discharge through the subsurface drainage system, in combination with the natural drainage, was more than sufficient to decrease and control soil salinity (Table 2.7). Over the three-year period the calculated field application efficiency was about 0.76. Like the water table, the discharge of the pipe drains showed great variations with time, being high just after irrigation, decreasing rapidly and often ceasing completely before the next irrigation. A strong relation was found between the discharge of the lateral drains and the crop cultivated in the drainage unit. Table 2.9 presents a summary of the measured drain discharges for the different crops. The analysis showed that peak flows occurred for a short time only. The highest flow rates were monitored during the rice season with a seasonal average of 1.3 mm d^{-1} and 90% (CF) of 2.4 mm d^{-1}. Rice, however, is a wet-foot crop which does not require the control of the water table by deep drains. Therefore, the drain discharges of rice cannot be considered a criterion for determining the drain spacing.

The seasonal average drain discharges of all other crops did not exceed 0.4 mm d^{-1}, the rate related to the maize crop. As the estimated maximum natural drainage rate in the area is of the order of 0.5 mm d^{-1}, a design rate of 0.9 mm d^{-1} should be considered if no natural drainage exists. This value is close to the design criterion of 1.0 mm d^{-1}. Nevertheless, as a result of the natural drainage in the area, the drain spacing should have been designed for about half this value. The 90% is quite arbitrary and additional research into the relation between high water tables and crop yield can refine this percentage. Table 2.9 shows that rice had the highest discharge, but again, rice is not the most critical crop because its yield is not affected by a high water table. So the critical rate is 1.2 mm d^{-1} being the 90% cumulative discharge of maize. Thus, instead of the currently used 4.0 mm d^{-1}, the design rate for drain pipe capacity can be reduced to 1.2 mm d^{-1}. For areas not subject to natural drainage, this rate should be increased to 1.7 mm d^{-1}. Several crops are usually cultivated at the same time in the catchment area of a collector drain [173]. Hence the discharge from a collector drain depends on the combination of crops grown in its catchment area.

Table 2.9   Drain discharges for the drainage units cultivated with the same crop.

Crop	Discharge (mm d^{-1})		
	Maximum	90% (CF)*	Seasonal average
Short *berseem***	4.3	0.2	0.2
Long *berseem***	6.7	0.8	0.3
Wheat	6.0	0.3	0.1
Cotton	2.4	0.3	0.1
Maize	4.1	1.2	0.4
Rice	4.8	2.4	1.3

*   CF = cumulative frequency of non-exceedance
**   refers to the duration of the growing season (see Figure 2.5)

Table 2.10   Collector drain discharges

Collector	Drained area	Discharge (mm d^{-1})		
(no.)	(ha)	Maximum	90% (CF)*	Seasonal average
I	52.7	2.0	1.3	0.6
II	20.2	3.9	1.8	0.6
III	17.8	3.9	1.4	0.5

Table 2.10 summarizes the statistical analysis of the discharges from the three collector drains in the Mashtul pilot area. Although relatively high peak discharges occur in the collector drains, they are not as high as for the pipe drains (Table 2.9). The peak discharges are smaller for the collectors with the biggest catchment area, showing an area reduction effect. The collector drain discharges vary slightly between the winter and summer seasons, but depend mainly on the cropping pattern [13]. Generally *berseem* (in winter), maize and rice (in summer) cause the higher drain discharges. The cumulative frequency in Table 2.10 shows that a drain discharge of less or equal than 1.8 mm d^{-1} prevails for 90% of the time. Hence, the design rate for collector drains in areas similar to Mashtul, with a rice intensity of maximum 60% and a natural drainage rate of 0.5 mm d^{-1}, can be reduced from 4 mm d^{-1} to 1.8 mm d^{-1}.

Overpressure in the subsurface drainage system

The monitoring programme of water levels in the collector system showed that the pipes were flowing full at discharges below the design rate [173]. Water levels above the invert level of the pipe occurred frequently in the upstream parts of the collector drains. Although this overpressure was mostly less than 0.15 m, pressures upto 0.50 m occurred. The overpressure can be attributed to a combination of the following effects [195]:

- irregularities in the alignment, the deviation from the average alignment varies between 0.05 and 0.15 m;
- obstructions (roots, sediment, reeds, etc.) in the collector drains, which reduced the effective cross-sectional area in some sections by approximately 35%.

Thus the design roughness factor is insufficient to take the currently used materials (concrete pipes with a joint at every 0.75 m and no collar) and construction methods (trenchers without laser control) into account. For the design of collector drains, Cavelaars recommended the use of the Manning equation with a roughness coefficient of 0.014 (K_m = 70), in combination with a safety factor of 100% for sandy soils and 67% for clay soils [49]. El-Atfy incorporated this safety factor, which is used to take the prevailing operational conditions and current quality of construction into account, in the roughness coefficient and recommended a value of 0.028 [69]. When this value of the roughness coefficient is used, no additional safety has to be incorporated in the other design factors (e.g. the design rate).

Conclusions

A drainage pilot area was constructed at Mashtul in the south-eastern part of the Nile Delta to verify the design criteria for subsurface drainage systems in this region. The monitoring programme showed that the crop yields increased significantly after installation of the subsurface drainage system. The increase was 10% for rice, 48% for *berseem*, 75% for maize and more than 130% for wheat. Part of the yield increase can be attributed to the decrease in soil salinity; the other part is the effect of improved water and air conditions in the root zone and improved agricultural inputs. The relation between crop yield and the depth of the water table showed that the optimum seasonal average depth of the water table midway between the drain is 0.80 m. A discharge of 0.4 mm d^{-1} is needed to drain the most critical crop (maize). Considering the natural drainage in the pilot area, which is estimated at 0.5 mm d^{-1}, the total drainage coefficient should be 0.9 mm d^{-1}. For areas like Mashtul, the most cost-effective way to obtain the above mentioned water table depth at the given discharge is to install drains at a depth between 1.20 to 1.40 m. For the determination of the capacity of the pipe drains a design rate of 1.2 mm d^{-1} is sufficient. For collector drains this design rate should be 1.8 mm d^{-1}, to take the high discharges from rice fields into account. For areas other than Mashtul, which are not subject to natural drainage, these rates should be increased by 0.5 mm d^{-1}. If the Manning equation is used for the design of the pipe system, a roughness coefficient (n) of 0.028 should be applied to take into account the current construction materials and methods. When this value of the roughness coefficient is used, no additional safety has to be incorporated in the other design factors (e.g. the design rate).

2.10 Water balance study in a drained area

The data collected in Mashtul pilot area during the monitoring programme was used to assess the overall water balance [175]. In general, the water balance in the root zone of an irrigated area reads [57; 267]:

$$I + P + G = E + R + \Delta W \qquad\qquad (\text{Eq 1})$$

where:

I	irrigation water (mm d^{-1})	
P	precipitation (mm d^{-1})	
G	rate of capillary rise from the saturated zone (mm d^{-1})	
E	evapotranspiration (mm d^{-1})	
R	percolation (mm d^{-1})	
ΔW	change in soil water storage in the root zone (mm d^{-1})	

and the corresponding salt balance:

$$I \times C_i + P \times C_p + G \times C_g = R \times C_r + \Delta Z \qquad\qquad (\text{Eq 2})$$

where

C	salt concentration (dS m^{-1})
i, p, g, r	suffix denoting irrigation, precipitation, groundwater and deep percolation
ΔZ	change in salt content in the root zone (kg ha^{-1} d^{-1})

It is assumed that $C_g = C_r$ and that the net deep percolation, being the difference between the total amount of deep percolation (R) and the amount of capillary rise (G), equals the sum of drainage through the subsurface drainage system (D) and the natural drainage (N).

In the field studies, not all components of the water and salt balance could be measured, i.e. the natural drainage and its salinity [2]. The natural drainage was calculated with the Darcy equation [40] for each crop season using piezometer readings and the corresponding depth of the water table (Table 2.11). The natural drainage varies per season and the 3-year average (0.3 mm d^{-1}) is in good agreement with groundwater studies conducted by RIGW that estimate the natural drainage in this area at about 0.6 mm d^{-1}, being located on the boundary between the zones of 0.0 to 0.5 mm d^{-1} and 0.5 to 1.0 mm d^{-1} [189] .

Another way to assess the natural drainage rate is to consider it as the closing factor of the water balance. Irrigation intervals and applications were measured to calculate the total amount of irrigation given to the various crops. The evapotranspiration was calculated with the modified Penman equation [76] using the meteorological data collected in the pilot area and the corresponding crop factors. If the water balance is considered over a 3-year period, so that it includes the complete crop rotation, the average natural rate is 0.5 mm d^{-1} (Table 2.12). The water balance does not fit between the individual cropping seasons. Beside inaccuracies in the estimations of individual components of the water balance, this is probably because the soil moisture changes between the cropping seasons ($\Delta W \neq 0$). The overall field application efficiency E_a, which is defined as the ratio between the volume of water needed, and made available, for evapotranspiration by the crop to avoid water stress in the plants throughout the growing cycle [43], is about 0.75.

The corresponding change in soil salinity has been calculated using a spreadsheet model based on the method described by Van Hoorn and Van Alphen [267] (Table 2.13). Over the seasons the soil salinity (EC_e) varies between 1.4 and 2.0 dS, well below the critical values for the crops grown in the area (Table 2.7). Thus, there are possibilities to increase the irrigation efficiency and to reduce subsurface drainage rated by controlled drainage. The minimum amount of irrigation water needed to keep the soil salinity below these critical levels can be calculated with [267]:

$$I = (E - P)\frac{2\,EC_e}{2\,EC_e - EC_i} \qquad (\text{Eq 3})$$

and the corresponding percolation requirement with:

$$R^* = (E - P)\frac{EC_i}{2EC_e - EC_i} \qquad (\text{Eq 4})$$

The leaching fraction, which is defined as the ratio between the net deep percolation and the irrigation, can be calculated with:

$$LF = \frac{R^*}{I} = \frac{EC_i}{2EC_e} \qquad (\text{Eq 5})$$

Table 2.11   Calculation of the natural drainage per cropping using the Darcy equation (with a hydraulic conductivity of 2.5 mm d^{-1} and the depth of the clay layer is 8.0 m)

Crop	Period	Cropping season (days)	Depth of water table (m)	Piezometric head (m)	Natural drainage (mm d-1)
Berseem	Nov-Mar	151	0.87	1.55	0.2
Cotton	Apr-Oct	214	1.00	1.53	0.2
Wheat	Nov-May	212	0.92	1.59	0.2
Rice	Jun-Oct	153	-0.10	1.48	0.5
Berseem	Nov-May	212	0.83	1.59	0.2
Maize	Jun-Oct	153	0.86	1.48	0.2
Overall		1,095	0.73	1.54	0.3

Table 2.12   Seasonal water balance

Crop	cropping season (days)	E-P (mm d^{-1})	I (mm d^{-1})	D (mm d^{-1})	N (mm d^{-1})	E$_a$ (-)	LF (mm d^{-1})
Berseem	151	1.5	1.7	0.2	0.0	0.89	0.11
Cotton	214	3.4	3.3	0.1	-0.2	1.03	-0.03
Wheat	212	1.7	2.0	0.1	0.2	0.85	0.15
Rice	153	3.4	6.8	1.2	2.1	0.51	0.49
Berseem	212	2.4	2.3	0.3	-0.3	1.02	-0.02
Maize	153	2.6	5.0	0.4	2.0	0.53	0.47
3-year rotation	1095	2.5	3.4	0.3	0.5	0.75	0.25

The irrigation water requirements for optimal leaching, I$_{min}$ ($\approx$ 2,890 mm over the 3-year crop rotation) is about 25% less than the amount of irrigation water that was applied (3,680 mm) (Table 2.14). To illustrate the options for controlled drainage, the soil salinity has been calculated for two scenarios:

- Scenario I: Optimum leaching efficiency (LF = 0.05 and I calculated with Eq 3), as the majority of the leaching takes place during the rice season the subsurface drainage system can be closed during the growing seasons of the other crops, reducing the overall drain discharge;
- Scenario II: Controlled drainage with optimum level of standing water in rice fields (0.10 m). Controlled drainage can not completely stop the subsurface drainage (Section 2.7), thus extra irrigation water is required to maintain a optimum water level in the rice plots. Consequently, this option required higher overall irrigation gift (about 3,430 mm over the 3-year rotation), but still about 9% less than the traditional irrigation and drainage practices.

Both scenarios for controlled drainage result in slightly higher soil salinity levels, but still well below the critical values, clearly illustrating the potential to further reduce drainage outflows by controlled drainage (Figure 2.15).

Table 2.13   Soil salinity per cropping season calculated using a spreadsheet model based on the method described by Van Hoorn and Van Alphen [267]

General data	W_{fc} =	440 mm				$EC_e = 0.5\ EC_{fc}$	
				Cropping season			
Period	3 year - rotation	Berseem	Cotton	Wheat	Rice	Berseem	Maize
Days	1095	151	214	212	153	212	153
EC_e - critical		3.2	7.7	3.2	3.0	3.2	2.4
E mm d^{-1}	2.5	1.6	3.4	1.7	3.4	2.4	2.6
E mm	2,774	237	732	362	528	515	401
P mm	26	9	0	9	0	9	0
E-P mm	2,748	228	732	353	528	507	401
EC_i dS m^{-1}	0.34	0.34	0.34	0.34	0.34	0.34	0.34
Existing practice							
I mm d^{-1}	3.4	1.7	3.3	2.0	6.8	2.3	5.0
I mm	3,677	258	710	418	1,035	497	759
R* mm	929	30	-21	65	507	-10	358
R* mm d^{-1}	0.8	0.2	-0.1	0.3	3.3	0.0	2.3
Dr mm d^{-1}	0.3	0.2	0.1	0.1	1.2	0.3	0.4
Dr mm	369	24	30	12	190	57	55
N mm	560	5	-52	53	318	-67	303
N mm d^{-1}	0.5	0.0	-0.2	0.2	2.1	-0.3	2.0
Z1 kg ha^{-1}		575	624	978	1,001	565	756
Z2 kg ha^{-1}		624	978	1001	565	756	572
EC_e* dS m^{-1}	1.4	1.4	1.6	2	1.7	1.4	1.2
EC_m dS m^{-1}	1.3	1.4	1.5	1.0	1.3	1.0	23
Scenario I - Optimum leaching efficiency with controlled drainage							
I mm	2,975	265	750	370	560	590	440
I mm d^{-1}	2.7	1.8	3.5	1.7	3.7	2.8	2.9
R* mm	227	37	19	17	33	84	39
R* mm d^{-1}	0.6	1.5	0.6	1.4	0.2	1.5	0.7
Z1 kg ha^{-1}		800	823	1,043	1,129	1,233	653
Z2 kg ha^{-1}		823	1,043	1,129	1,233	653	744
EC_e* dS m^{-1}	2.2	1.9	2.1	2.4	2.7	2.3	1.5
Scenario II - Controlled drainage with standing water in rice fields							
I mm	3,425	260	740	360	1,035	590	440
I mm d^{-1}	23.7	10.7	24.3	30.0	5.5	10.3	8.0
R* mm	677	32	9	7	507	84	39
R* mm d^{-1}	0.6	0.2	0.0	0.0	3.3	0.4	0.3
Z1 kg ha^{-1}		780	811	1,047	1,152	612	694
Z2 kg ha^{-1}		811	1,047	1,152	612	694	781
EC_e* dS m^{-1}	1.7	1.8	1.4	2.4	1.9	1.4	1.6

Table 2.14   Irrigation water requirements for optimal leaching

Crop	EC_e (dS m^{-1})	I-applied (mm)	I_{min} (mm)	LF (-)
Short *Berseem*	3.2	258	241	0.05
Cotton	7.7	710	748	0.02
Wheat	3.2	418	373	0.05
Rice	3.0	1,035	559	0.06
Long *Berseem*	3.2	497	535	0.05
Maize	2.4	759	432	0.07
3-year rotation		3,677	2,887	0.05

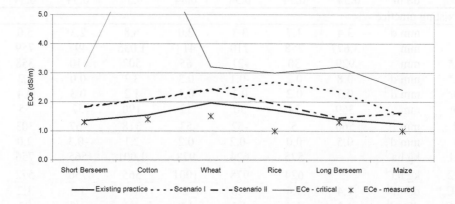

Figure 2.15   Soil salinity (EC$_e$) of the root zone for the existing irrigation practices, controlled drainage in combination with optimum leaching efficiency (scenario I) and with maintaining a layer of standing water in rice fields (scenario II).

To verify these results the model SALTMOD was used. SALTMOD is a computer program for the prediction of the salinity of soil moisture, groundwater and water table depth and drain discharges in irrigated agricultural lands [157]. The model was calibrated using the seasonal average depth of the water table. The simulated drain discharges varied between 0.2 and 0.6 mm d^{-1} in summer and between 0.2 and 0.4 mm d^{-1} in winter and the natural drainage was in the range 0 and 0.7 mm d^{-1} [158]. Although, these rates still have a large variation, these simulations are in the same order of magnitude as the results of the field measurements.

3   Subsurface drainage practices in India

3.1   History of irrigation and drainage in India

Although India, in recent years, is emerging as an industrial nation, agriculture remains a key sector in India's economy, contributing about 35% of the Gross Domestic Product and employing 72% of its adult population [105]. Agriculture in India is diverse: the cropping patterns and cultivation methods are largely determined by the diverse geographical and climatic conditions and the structure of the agricultural organization, rural economy and rural society is the result of very diverse historical conditions [211]. Development plans of the Central and State Governments give priority to alleviate poverty and to create employment, particularly in rural areas, but annual agricultural growth has been modest at 2.6% per annum over the last 25 years [103]. The average Indian farmer is a smallholder owing less than one hectare of cultivable land, harvesting one crop a year and striving to harvest a second crop [173]. Agriculture depends largely on the monsoon; rains, however, are unevenly distributed in time and space. To sustain agricultural production against these vagaries of rainfall, irrigation has been introduced in about 57 Mha, covering about 34% of the total arable land (Table 1.1).

Irrigation in India is believed to be as old as the history of agriculture. Since ancient times, different types of irrigation systems (and agricultural practices) were used in different parts of the country, depending on the soil, hydrological and climatic conditions [173]. The most well-known system, tank irrigation, is still widely used today. Tanks are man-made or natural reservoirs in which river water and rainfall is stored during the *Kharif* (monsoon season from July to October) to irrigate crops up to the *Rabi* (post-monsoon or winter season from October to March). Canal irrigation, which started during the British reign in the 18th century, greatly expanded after independence [91]. Canal irrigation systems are divided in major (with command areas > 10,000 ha), medium (between 2,000 and 10,000 ha) and minor (< 2,000 ha). Groundwater is the third source of irrigation water. In most canal command areas, conjunctive use is made of surface and groundwater as the supply of canal water is often insufficient. It is estimated that today, about 76% of the net irrigation comes from groundwater, with conjunctive use this percentage increases to over 87% [173]. At present, 57 Mha of the agricultural land is irrigated, about one third of the cropped area and 41% of the potential (Table 3.1). The overall utilization of the water resources is about 65%, thus irrigation is by far the biggest consumer.

Most irrigation systems in India are not designed to cover the full crop water requirement but on the principle that the available water is spread over a large area, the so-called 'protective' irrigation [117]. The idea is to reach as many farmers as possible and to protect them against crop failure and famine.

The introduction of irrigated agriculture, however, has resulted in the development of the twin problem of waterlogging and soil salinization. Considerable areas have either gone out of production or are experiencing reduced yield. With the misconception, that the more they irrigate, the more yield they will get, farmers apply huge quantities of canal water: e.g. in Segwa, one of the study areas discussed in Section 3.5, the actual supply of 2,924 mm y^{-1} by far exceeds the crop water requirement of 1,912 mm y^{-1} (Section 3.7).

Table 3.1   Potential and implemented irrigation [159].

	Irrigation potential (10^6 ha)	Utilized (%)
Major and medium irrigation	58.5	56
Minor irrigation	17.4	71
Groundwater irrigation	64.1	71
Total	139.9	57

Furthermore, the introduction of canal irrigation not only brings the much-needed water, but also imports salts as irrigation water contains considerable amounts of salt. In Segwa, the canal water has a salinity of 0.3 dS m^{-1}, thus an irrigation gift of 1,910 mm y^{-1} will add 3.7 t ha^{-1} of salts to the soil profile.

The adverse effects of excess water were already known in ancient times. Gupta [90] refers to a hymn in the Narda Smirty, a Hindu Epic, in which the sage Narada said that '*No grain is ever produced without water, but too much water tends to spoil the grains and inundation is as injurious to growth as is the dearth of water*'. Thus, although the need for drainage was realized long ago, drainage has never received much importance. Even today, irrigation projects are planned with the hope that either drainage might not be needed or funds will be made available once the project starts yielding revenues. As a result, considerable areas have either gone out of production or are experiencing reduced yield due to waterlogging and/or salinity problems. In only about 2.5 Mha of the affected lands some sort of drainage system has been installed [105]. Subsurface drainage was introduced only recently. Thus far, only about 21,000 ha waterlogged saline land in canal irrigation commands have been provided with subsurface drainage systems, the majority of which is less than 10 years old [90]. The bulk of the coverage is in Rajasthan (15,000 ha), Haryana (2,000 ha), Maharashtra (2,000 ha) and Karnataka (2,000 ha). The subsurface drainage experiences in the country are mainly in these four states (Figure 3.1).

India experiences a wide range of climatic and physiographic conditions and a correspondingly wide range of waterlogging and soil salinity problems. These problems are broadly classified into three groups [91]:

- Rainfed induced waterlogging is found naturally in imperfectly drained land in much of the country during the monsoon season, with the exception of the arid parts of Gujarat and Rajasthan. The rainfall is intense during the 3 - 4 monsoon months and surface drainage is required to overcome waterlogging problems;
- Natural salinity occurs in various locations in the semi-arid parts of the north-western and western part of the Gangetic plain under the prevailing hydrologic and geochemical conditions of the land. Natural salinity is also found in the plains and deltaic areas along the coast;
- Irrigation-induced waterlogging and salinity is a relatively recent feature that developed in the late 19th century when large-scale canal irrigation was introduced. These problems are found in different command areas throughout the country, either in the form of waterlogging only, or a combination of waterlogging and soil salinity.

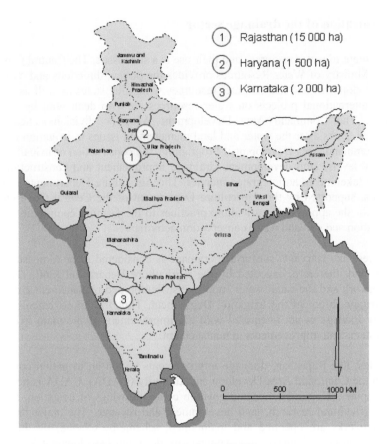

Figure 3.1   Subsurface drainage projects in India [150]

It is estimated that some 8.4 Mha is affected by soil salinity and alkalinity, of which about 5.5 Mha is also waterlogged, mainly in the irrigation canal commands and 2.5 Mha in the coastal areas. Even though some surface drainage improvements have been undertaken in rainfed and irrigation-induced waterlogged areas, the drainage requirements of much of the agricultural lands are yet to receive much attention. The participants of the 8th ICID International Drainage Workshop with the theme '*Role of Drainage and Challenges in 21st Century*', organised in New Delhi in 2000, concluded that drainage was often considered as an adjunct to irrigation and was not considered as a discipline in its own right [100]. The improvement of drainage in India is largely planned and implemented as a flood control measure. As a result most of the attention has been on the construction of flood protection embankments along the rivers and major surface drains. Only in recent years, on-farm drainage has started to receive some attention in irrigation command areas.

Research for the control of waterlogging and soil salinity received a big boost under the Canadian-aided Rajasthan Agricultural Drainage Research project (RAJAD) [182], the Netherlands aided Haryana Operational Pilot Project (HOPP) [260] and the Indo-Dutch Network Project (IDNP) [103]. The necessary research developments for large-scale expansion of subsurface drainage have emerged from these projects.

3.2 Organization of the drainage sector

The development of water resources and their use is a state issue. The Central Government through the Ministry of Water Resources provides the policy, directions and expertise for planning, development and use of water resources within the states as well as among the states [89]. International projects on water resources are also dealt with by the Central Government. The Command Area Development Wing (≈ divisions), headed by a Commissioner, looks after the water and land management issues in irrigation commands. Under the Command Area Development Programme, the construction of irrigation systems up to tertiary levels, land development, drainage improvement and construction of road networks are taken up in an integrated manner. The Agriculture Departments of the State Governments, supported by the Agriculture Department of the Central Government, are responsible for the appropriate agriculture practices to the farming community as well as land reclamation and soil and water conservation activities.

In some states, like Haryana, subsurface drainage is considered to be a measure for land reclamation and, therefore, executed by the Agriculture Department [260]. The surface drainage network and the canal water supply, distribution and management are, however, under the responsibility of the Irrigation Department. Thus the improvement of on-farm (subsurface) drainage is not integrated with the improvement of the main irrigation and drainage systems and improvements in management.

In other states, like Rajasthan, drainage improvements within an irrigation command are entrusted to the Command Area Development Authority (CADA). CADA is headed by an Area Development Commissioner under whom come the wings of irrigation, agricultural extension, agricultural research, land development and revenue. The major drainage and irrigation networks are operated and maintained, scheme-by-scheme, by the Irrigation Wing and the on-farm drainage improvements by the Land Development Wing. Outside the command areas, the improvement of drainage is the responsibility of the Irrigation Department. Thus there is no integrated and centralized organization for improving drainage and the policy and practices are not in balance as they vary with each department.

Research on water management and drainage is conducted under the Indian Council of Agricultural Research (ICAR) through its own specialized research organizations. One of these specialized research organizations, the Central Soil Salinity Research Institute (CSSRI), established in 1969 at Karnal, is conducting research to combat waterlogging and salinity. Since 1985, Alterra-ILRI, through bilateral projects, has cooperated with CSSRI on these research activities [103]. The research mainly focuses on crop drainage requirements and land reclamation practices. Construction and O&M have not yet received much attention.

In 2004, the central government, under the restructured Command Area Development and Water Management Programme has revised the funding norms for subsurface drainage systems. The funding, Rs 40,000 (about € 635) per ha, is shared by the central government, state governments and farmers in the ratio of 50 : 40 : 10 [86].

3.3 Design principles

Layout

Composite subsurface drainage systems are used [150]. The field pipe drains connect to a collector pipe drain that discharges by gravity in an open drain or into a sump. A sump serves around 50 ha, an area of approximately 40 - 50 farmers, being the maximum size of group that can organise themselves. From the sump, the drainage water is lifted by pumping and disposed of into the open main drainage system.

Design criteria

For the prevailing conditions in India, subsurface drainage systems with rather deep drains, i.e. drain depth > 1.75 m, are recommended [90], although the applied drain depth/spacing combinations vary considerably between the various agro-climatic regions (Table 3.2). The recommendation is based on the critical depth concept, i.e. to avoid secondary salinization caused by the upward flux of water once the water table rises to a depth of 2 - 3 m [91]. These deep drains have their drawbacks. Firstly, the deeper the drain, the higher the installation cost. Secondly, deep drains can only economically be installed by mechanical construction practices, ignoring the huge employment needs of the rural poor. Thirdly, deep drains lower the water table during the irrigation season. These lower water tables reduce the rate of capillary rise and thus increase the burden on the already poorly performing irrigation systems.

Research in countries with similar conditions, i.e. Egypt and Pakistan, indicates that shallower drains can maintain salinity levels within safe limits for crop production [14]. Research conducted in various agro-climatic regions in India also suggested that the drain depth can be reduced.

The design discharge is based on salinity control under the assumption that monsoon rainfall is adequate for salt removal. The design rates vary between 1.5 and 2 mm d^{-1}. Under the prevailing conditions in India, this results in drain spacings varying between 40 m and 60 m. In sandy areas the spacing can be more than 100 m (Section 3.5).

Table 3.2   Recommended drain depth-spacing combinations for various agro-climatic regions in India

Agro-climatic region	Drain depth (m)	Drain spacing (m)	References
Semi-arid coastal plains of Andhra Pradesh	1.4	10 - 15	[185]
Semi-arid Trans-Gangetic plains of Haryana	1.4 - 1.75	60 - 100	[187; 260]
Humid coastal plains of Kerala	1.0	30	[138]
Semi-arid plains of Gujarat	1.0	20 - 40	[162]
Arid lands of Rajasthan	1.0 - 1.5	30 - 60	[163; 182]
Sub-humid regions of the lower Gangetic plains in West Bengal	1.75	15 - 45	[185]

3.4   Installation practices

Planning

As there are no centralized organizations in India to diagnose, monitor and implement drainage measures there is also no systematic planning to address the drainage problems. The drainage improvements are limited to waterlogged areas in the form of open drainage and subsurface drainage for waterlogged saline areas. The monitoring activities, identification and implementation are spread over several departments.

Pipe and envelope materials

Bell-mouthed clay pipes with a row of eight perforations on the underside, 60 cm in length and 100 mm diameter for field drains and 150 mm diameter for collectors, were used in South India in the 1970s and 1980s [150]. The material cost was three times the cost of installation. Furthermore, the performance of drainage systems was severely affected by the displacement of the pipes and choking of drain lines [186]. In the 1980s. concrete pipes were used for collector and field drains, for example in the Sampla area in Haryana [185]. The production of corrugated PVC drain pipes only commenced in the early 1990s and then only up to 100 mm in diameter conforming to *Deutsches Institut für Normung* (DIN, German Institute for Standardization) specifications. These corrugated PVC pipes were used for field drains and reinforced cement, concrete or PVC rigid pipes for collector drains. In the mid-1990s under the RAJAD project, corrugated PVC drain pipes of seven sizes became available, ranging in diameter from 80 mm to 450 mm [182].

Both granular and synthetic materials are used for envelopes. In areas where clay pipes were used, sand blinding, i.e. backfill with sand around the pipe, was practised. A considerable amount of testing of envelope materials has been done in sandy loam soils under the HOPP and IDNP projects and in clay soils under the RAJAD Project. Field investigations under RAJAD and laboratory investigations at the CSSRI showed the soil texture (clay percentage) and sodium absorption ratio (SAR) to be the significant determinants of the need for an envelope. The criteria for deciding on the need for an envelope are [126]:

- clay > 40%   Envelope not required
- clay 30 – 40% and SAR > 16   Envelope required
- clay < 30%   Envelope required

The specifications for the envelope materials adopted by respectively the HOPP and RAJAD project were as follows:

- *Sandy soils*: For sandy loam soils, non-woven polypropylene materials with a minimum thickness of 3 mm weighing 300 g/m^2 or more and with a O_{90} between 350 and 550 microns are recommended [260]. The envelope should be strong enough for manual or machine wrapping and for transport and installation by hand or machine;
- *Clay soils*: For clay soils, non-woven polypropylene material with a minimum thickness of 0.9 mm when compressed and a weight of at least of 240 gm/m^2 are recommended [182]. The permeability of the envelope should be at least 20 m/day and 95% of the openings of the envelope material should be smaller than 150 microns but not smaller than 100 microns. The tensile strength of the envelope should be at least 360 N.

Installation

Large-scale subsurface drainage installation is relatively new in India. International contractors imported the drainage machinery and undertook the bulk of the installation under the RAJAD project. For HOPP, the Netherlands Government provided a trencher. For small-scale projects, semi-mechanical methods (using an excavator) and manual installation methods are used (Section 3.6).

3.5   Disposal of the drainage effluent

Disposal options very much depend on the region. In landlocked parts of the country (e.g. West Rajasthan) disposal of drainage water is a serious problem that has not yet been solved. But also in the upper reaches of the main rivers (Ganges, Krishna and Govadari) disposal of drainage water in the *Rabi* and *Zaid* (summer season) is problematic as river flows are generally low and the drainage effluent adversely affects the salinity of the river water. Reuse of drainage water is an option for Northwest India as it could supplement scarce irrigation water supplies and also help to alleviate disposal problems [254]. Field experiments conducted for 3 - 7 years at the experimental station of CSSRI, at Karnal, Haryana, shows that reuse can minimize disposal needs [223]. Controlled drainage, in combination with reuse, can meet part of the crop water requirements in times of water scarcity and thus reduces the pumping needs.

3.6   Lessons learned in famers' fields [7]

Introduction

To develop location-specific guidelines for subsurface drainage, the Governments of India and the Netherlands jointly initiated the Indo-Dutch Network Operational Research Project on Drainage and Water Management for Control of Salinity and Waterlogging in Canal Commands' (IDNP) (1995-2002). The recommendations and strategies developed by this project are presented in this section [103].

Materials and methods

Six pilot areas in farmers' fields, one experimental plot and one large-scale monitoring site were established in those agro-climate regions where canal irrigation is most important, i.e. (Figure 3.2):

- Islampur/Devapur in the Southern Plateau and Hills of Karnataka;
- Konanki and Uppugunduru in the East Coast Plains and Hills of Andhra Pradesh;
- Lakhuwali in the Western Desert Region of Rajasthan;
- Segwa and Sisodra in the Plains and Hills of Gujarat;
- Gohana and Sampla in the Trans-Gangetic Plains of Haryana.

---

[7] Published as: Ritzema, H.P., Satyanarayana, T.V., Raman, S., Boonstra, J., 2008. Subsurface drainage to combat waterlogging and salinity in irrigated lands in India: lessons learned in farmers' fields. *Agricultural Water Management*, **95**: 179 – 189. doi: 10.106/j.agwat.2007.09.012

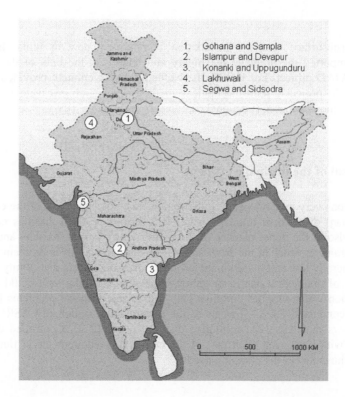

Figure 3.2 Location of the six pilot areas, experimental plot and the large-scale monitoring
         site.

Islampur/Devapur

Islampur/Devapur pilot area is located in the Upper Krishna Project (16°08'N and
75°37'E). The climate is semi-arid tropical monsoon with a mean annual rainfall of 768
mm and a potential evaporation of 2,180 mm. The area is irrigated with good quality canal
water and the main crops are rice, cotton, chillies, wheat during *Kharif* (monsoon season
from July to October) and sorghum during *Rabi* (post-monsoon or winter season from
October to March). During the *Zaid* (summer season from March to June) fields are
prepared for the *Kharif* crops. Most farmers are illiterate and poor (Table 3.3): 23% of the
farmers have land holdings smaller than 1 ha. Although the quality of the irrigation water
is good ($EC_i = 0.6$ dS m^{-1}) waterlogging and salinity problems occur. The causes are the
poorly drainable black soils, seepage from the canal network, lack of land development,
inefficient irrigation practices and inadequate drainage.

Konanki and Uppugunduru

Konanki and Uppugunduru pilot areas are located in respectively the Nagarjunasagar
Project Right Canal Command (15°48'N and 80°21'E) and the Krishna Western Delta
(15°45'N and 80°19'E). The climate is predominantly semi-arid to arid with a mean annual
rainfall of 768 and 844 mm for respectively Konanki and Uppugunduru and both with a
potential evaporation of about 1,600 mm.

Table 3.3   Socio-economic characteristics of the farmers in the pilot areas

Area	No. of farmers	Average		M/ F	Education level[a] (%)			
		family size (no.)	farm size (ha)		I	P	S	D
Islampur	59	7.1	3.2	M/ F	56	← 39 →		5
Konanki	30	4	0.54	M	36	20	40	4
				F	58	11	24	7
Uppugunduru	41	4	0.49	M	32	32	27	9
				F	54	24	20	2
Lakhuwali	36	8.8	2.3	M	55	26	18	1
Segwa	52		3.6	M	0	40	52	8
				F	0	34	52	14
Sisodra	47		2.1	M	13	68	15	4
				F	19	68	13	0

[a]   M-male; F=Female; I= Illiterate; P = Primary level (Standard 1 to 7);
S = Secondary level (Standard 8 to 12) and D = Diploma and above

The main crop in Konanki and Uppugunduru is rice. Soils are sandy clay to clay loam. The average farm size is about 0.5 ha and the education level of the farmers is low (Table 3.3). The majority of the farmers are relatively poor. Although, the quality of the irrigation water is good (EC_i = 0.6 dS m^{-1}) yields are low due to waterlogging and salinity problems.

Lakhuwali

Lakhuwali pilot area is located in the Indira Gandhi Nahar Pariyojana Command in north-west Rajasthan (29°30'N and 74°24'E). The climate is arid to semi-arid with a mean annual rainfall of 297 mm and a potential evaporation of 1,560 mm. The area is irrigated with good quality irrigation water from the Indira Gandhi Main Canal (EC_i = 0.3 dS m^{-1}) and the main crops are cotton and wheat. The farmers are poor and illiterate with an average farm size of 2.3 ha (Table 3.3). The soil is a coarse textured sandy soil; a calcareous layer at shallow depth results in the formation of a perched water table. The area has no natural drainage outlet and after the introduction of irrigation, the water table rose at a rate of about 1 m y^{-1}, resulting in complete waterlogged conditions.

Segwa and Sisodra

Segwa and Sisodra pilot areas are located in the Ukai-Kakrapar Command in south Gujarat. Segwa, situated in the middle branch of the command (21°05'N and 73°00'E) has a mean annual rainfall of 1,500 mm and a potential evaporation of 1,770 mm. Sisodra, situated in the lower part of the command (21°11'N and 72°55'E) receives less rainfall (850 mm y^{-1}), but has approximately the same potential evaporation (1,670 mm y^{-1}). Segwa is partly irrigated with good quality canal water (EC_i = 0.3 dS m^{-1}) and partly with poor quality groundwater (EC_i = 2.45 dS m^{-1}). Sisodra completely depends on canal water, which is in short supply, so the drainage water is reused. Soils are heavy (black cotton, clay content > 45%) and the main crops are sugarcane and rice. Farmers in Segwa have a relatively high education level (Table 3.3) and, with an average farm size of 3.6 ha, well-to-do: 62% of the households have an yearly income of more than € 4000. In Sisodra,

farmers are poorer, only 11% of the households have an yearly income of more than €
4000. They have a lower education level and smaller farm size (Table 3.3). Both Segwa
and Sisodra have good outlet conditions, but suffer from waterlogging and salinity
(including alkalinity), because of heavy rainfall during monsoon, inefficient irrigation
methods, and poor permeability of the heavy clay soils.

Gohana and Sampla

The Gohana monitoring site is located in the Western Jamuna Canal command in Haryana
(29° to 29°10'N and 76° 42'to 76° 52'E). The climate is semi-arid monsoon with a mean
annual rainfall of 550 mm and a potential evaporation of 1650 mm. The main crops are
rice, sorghum, pearl millet and sugarcane during *Kharif* and wheat, mustard, barley and
berseem during *Rabi*. Gohana is part of the Haryana Operational Pilot Project, a project to
install subsurface drainage in 2000 ha [260]. The main problems in the area are saline
groundwater, high water tables and surface water stagnation during *Kharif*. The
experimental site Sampla is also located in the Western Jamuna Canal Command in
Haryana, just south of Gohana (29°08'N and 76°36'E). Sampla, constructed in 1984, was
used to study the long-term effects of subsurface drainage.

The research was conducted in farmers' fields: the drainage infrastructure was specially
designed and implemented for the research needs. Two to three years before the
installation of the subsurface drainage system, pre-drainage investigations and base-line
surveys were conducted to collected data on the climate (rainfall and evaporation), soils
(texture, salinity status, hydraulic conductivity, etc.), irrigation and drainage practices,
water quality, crops and the socio-economic status of the farmers (farm size, education,
labour requirements, income, farming practices, etc.). Regular meetings with farmers were
organized to discuss their involvement in the pre-drainage investigations and their
preferences and willingness to be involved in the construction and O&M of the subsurface
drainage systems. Four types of subsurface drainage systems, with different depth-spacing
combinations, were installed (Table 3.4):
- composite pipe drainage in which both field and collector drains are buried pipes;
- singular pipe drainage in which the field drains are buried pipes that discharge in
  an open collector drain;
- composite open drainage in which both field and collector drains are open drains;
- singular open drainage with only open field drains discharging directly in an open
  collector drain.

Corrugated PVC pipes with a diameter of 80 mm were used for the field drains and rigid
PVC pipes with diameters of 160 and 180 mm for the collector drains. In the lighter soils,
the pipes were pre-wrapped with synthetic envelopes. The open drains have a bottom
width varying between 0.40 and 0.50 m and a depth varying between 0.70 m (for the field
drains) to 1.00 m (for the collector drains). For the construction, whenever possible, local
available materials (pipes and envelopes) and equipment were used.

Table 3.4   Main characteristics of the subsurface drainage systems installed in the pilot
              areas

Pilot area	Type	Spacing (L) (m)	Depth (D) (m)	Installation cost * (€ ha⁻¹)
Konanki	Composite open gravity	100	1	89
Sisodra	Composite open gravity	60	0.70-0.80	144
Uppugunduru	Composite open pumped	50	1.00-1.20	191
Sisodra	Composite open gravity	30	0.70-0.80	268
Islampur	Singular open gravity	50	1.00-1.10	214
Segwa	Singular pipe gravity	45	0.90-1.20	432
Islampur	Singular pipe gravity	50	1.00-1.20	484
Konanki	Composite pipe gravity	60	0.90-1.10	452
Segwa	Composite pipe gravity	45	0.90	507
Lakhuwali	Composite pipe pumped	150	1.20	522
Uppugunduru	Composite pipe pumped	60	1.20-1.35	529
Sampla	Composite pipe pumped	50	1.20-1.50	554
Islampur	Composite pipe gravity	30-50	1.10-1.20	596
Uppugunduru	Composite pipe pumped	45	1.20-1.35	703
Konanki	Composite pipe gravity	30	0.90-1.10	753
Gohana	Composite pipe pumped	67	1.60	770
Uppugunduru	Composite pipe pumped	30	1.20-1.35	939

* 2002 prices (€ 1.00 = US$ 1.00)

The day-to-day farm management practices were fully controlled by the farmers. The research team only advised the farmers how to operate and manage the newly installed drainage systems. Next to the installation of drainage systems, location-specific measures to combat waterlogging and salinity were introduced, e.g. gypsum applications, lining of field irrigation canals, land levelling, use of organic manure, cultivation of green manure crops like *Dhaincha* and salt-tolerant rice varieties. To compare the pre- and post-drainage conditions in drained and non-drained (control) areas, a monitoring programme started two years before the installation of the subsurface drainage systems. Groundwater levels (daily measurements in 2 m deep observation wells placed at regular distances perpendicular to the field drains), drain discharges (daily measurements of drain pipe discharges and frequent flow measurements in the open drains), soils salinities and crop yields (twice per year) were collected. Furthermore, every year socioeconomic surveys were conducted to assess the cost and benefits of subsurface drainage and the farmers' attitude to the introduction of subsurface drainage systems in their fields.

Results and discussion

Installation

In India, with its huge population depending on agriculture, labour is abundant. Therefore two installation methods were tested, i.e.: (i) manual installation and (ii) a combination of manual and mechanical installation. For the second method, local available hydraulic excavators were used to dig the drain trenches up to 10-15 cm above the design bed level.

All other activities such as levelling, grading, envelope wrapping, pipe-laying, and backfill were done manually. The labour requirements per 100 m of drain line were:
- preparatory activities, e.g. setting out alignments and levels: 1 to 2 days;
- supporting activities, e.g. levelling, grading, smoothing the trench bottom, pipe-laying and backfill: 5 to 11 days;
- excavation of trenches (manual method only): 30 – 33 days.

The installation costs of subsurface drainage systems with different spacing and envelope material were recorded (Table 3.4). The overall costs of the manual installed systems (on average € 495/ha) were 27% higher than the costs of the semi-mechanical installed systems (€ 385/ha). For the pipe drainage systems, the cost of the materials, i.e. pipes and envelopes, accounted for about three-quarters of the total installation costs (Figure 3.3). This explains the big difference in cost between the pipe and open drainage systems.

Effects of drainage on crop yield, soil salinity and water table

The monitoring programmes, which continued two or three years after the installation of the subsurface drainage systems, clearly show that the installation of subsurface drainage systems resulted in higher crop yields (Figure 3.4), although there were differences between the various areas (Table 3.5). Overall, the yield increased from 2.5 to 4.3 t ha^{-1} for rice, from 0.7 to 1.2 t ha^{-1} for cotton, from 66 to 101 t ha^{-1} for sugarcane and from 1.4 to 2.8 t ha^{-1} for wheat. To assess how much of the increase in yield can be attributed to drainage, yields from fields with different drain depth/spacing combinations were compared with non-drained (control) areas (Table 3.6). The yield increases in the drained fields were significantly higher than the increases in the non-drained fields. These higher crop yields are the result of the two direct effects of drainage, i.e. in the drained fields, water tables were 25% deeper and the soil salinity decreased with 50%.

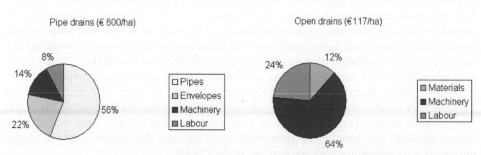

Figure 3.3   Cost components of the installation of open and pipe drainage systems (average costs of installation in Konanki and Uppugunduru pilot areas).

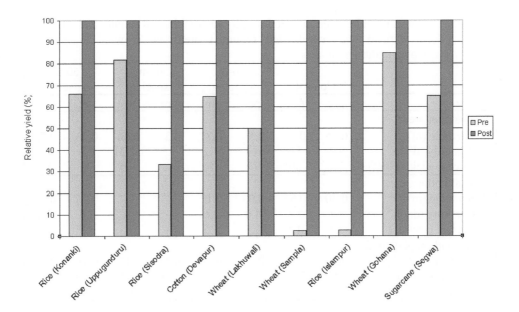

Figure 3.4   Pre- and post yield of the major crops in the drainage pilot areas.

Table 3.5   Results of the monitoring programme in the pilot areas

Pilot area	L	D	Crop	Water table (m)		Soil salinity (dS m^{-1})		Crop yield (t ha^{-1})	
	(m)	(m)		Pre[a]	Post	Pre	Post	Pre	Post
Konanki	30	0.900	rice	0.03	0.12	4.0	2.6	3.5	5.5
Konanki	60	0.90	rice	0.24	0.32	7.5	3.0	3.9	5.7
Uppugunduru	30	1.20	rice	0.50	0.75	7.7	2.6	4.5	5.6
Uppugunduru	45	1.20	rice	0.44	0.67	3.9	2.2	4.5	5.8
Uppugunduru	60	1.20	rice	0.46	0.53	3.5	1.5	4.5	5.1
Segwa	30	0.90	sc[b]	0.56	0.74	6.4	1.2	78.0	115.0
Segwa	45	0.90	sc	0.32	0.46	1.5	1.2	42.0	105.0
Segwa	60	0.90	sc	0.53	0.72	6.0	1.1	78.0	84.0
Sisodra	60	0.80	rice	0.55	0.88	16.3	12.3	0.6	1.8
Lakhuwali	150	1.20	wheat	0.30	1.00	8.2	8.3	1.4	2.8
Sampla	75	1.75	wheat	--	--	49.5	4.4	fallow	4.2
Islampur - open	50	1.00	rice	0.09	0.47	7.0	9.2	fallow	3.7
Islampur - pipe	50	1.00	rice	0.68	0.69	7.5	5.5	2.0	3.3
Islampur - pipe	30	1.10	rice	0.78	0.89	14.3	8.6	1.9	3.1
Islampur - pipe	30	1.10	cotton	0.78	0.89	14.3	8.6	0.7	1.2
Islampur - pipe	50	1.10	cotton	0.90	0.83	12.2	7.7	0.7	1.1
Gohana	60	1.70	wheat	--	--	7.1	4.6	3.1	3.6

[a] pre- and post: before and after the installation of the subsurface drainage system
[b] sc = sugarcane

Table 3.6   Comparison between the depth of the water table, soil salinity and crop yield
             in drained and non-drained (control) areas in Segwa, Konanki and
             Uppugunduru pilot areas

	Control		30-m spacing		45-m spacing		60-m spacing	
	pre	post	pre	post	pre	post	pre	post
Segwa pilot area:								
Water table (m)	0.35	0.35	0.53	0.74	0.32	0.46	0.53	0.72
Soil salinity (dS m^{-1})	3.3	5.0	6.4	1.2	1.5	1.2	6.0	1.1
Sugarcane yield (t ha^{-1})	75	80	78	115	42	105	78	84
Konanki pilot area:								
Water table (m)	0.40	0.56	0.03	0.12	--	--	0.24	0.32
Soil salinity (dS m^{-1})	3.1	3.1	4.0	2.6	--	--	7.5	3.0
Rice yield (t ha^{-1})	3.1	4.1	3.5	5.5	--	--	3.9	5.7
Uppugunduru pilot area:								
Water table (m)	0.54	0.70	0.50	0.75	0.44	0.67	0.46	0.53
Soil salinity (dS m^{-1})	28.0	25.5	4.7	2.6	3.9	2.2	3.5	1.5
Rice yield (t ha^{-1})	3.2	4.7	4.5	5.6	4.5	5.8	4.5	5.1

The SALTMOD model was used to predict long-term effects on soil salinity and the depth
to water table [244]. For calibration, the root zone salinity was simulated and compared
with observed data. The corresponding leaching efficiency was about 0.6 and the
simulations confirmed that the root zone soil water salinity had decreased: from an initial
value of 11.5 dS m^{-1} to 6.0 dS m^{-1} two years after the installation of the drainage system
(Figure 3.5). The model predicted the that root zone water salinity will further decrease to
about 2 to 3 dS m^{-1}. The model was also used to simulate different drain depth scenarios:
simulations show that deepening the drains from the present depth of 1.0 m to 1.4 m will
not further reduce the root zone salinity (Figure 3.6).

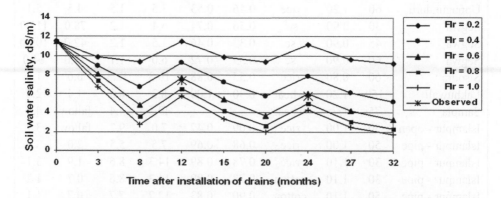

Figure 3.5   Effect of leaching efficiencies (Flr) on the root zone salinity in Konanki pilot
             area [101].

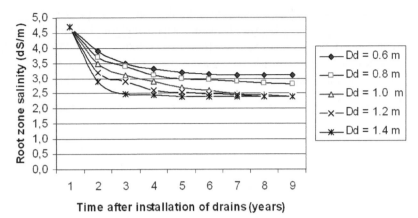

Figure 3.6   Simulated salinity in the root zone for different drain depths Dd [244].

Drain discharges

The 'hidden' cost of irrigation is a drain discharge with a high salinity. In Segwa, an area characterized by rather high annual rainfall (on average 1,500 mm y^{-1}) and an abundance of irrigation water, drain discharges are high (between 1.9 to 2.4 mm d^{-1}), but the salinity of the drainage water is relatively low (between 1.3 to 2.4 dS m^{-1}). In Konanki and Uppugunduru, both located in the tail end of their irrigation command and receiving less rainfall (respectively 768 and 844 mm y^{-1}), the discharges are lower (around 1.0 mm d^{-1}), but the salinity of the drainage water is higher (between 1.8 to 8.2 dS m^{-1}).

Envelope requirements

For areas with heavy clay soils, like Sega and Sisodra, there is no need for an envelope material: pipe drains with and without envelope performed equally effectively in these soils. Excavation programmes, conducted one or two years after installation of the pipe drains, showed only traces of sedimentation. For the lighter soils, non-woven polypropylene filter materials with a 0_{90}-value between 300 and 450 μm performed best, although they are rather expensive [126].

Costs and benefits

The introduction of the subsurface drainage resulted in an increase in both crop yield and cropping intensity and a shift from low to high value crops, e.g. black gram. In addition, the introduction of subsurface drainage reduced the workload of both men and women. For example, in Uppugunduru, a 20% reduction in labour input in rice cultivation was observed: men needed less time for flushing and women for (re-)planting and weeding (Figure 3.7). Overall, the introduction of subsurface drainage proved to be beneficial: for the most expensive (composite pipe drainage) systems, benefit-costs ratios vary between 1.2 to 3.2, internal rates of return between 20 to 58%, and pay-back periods between 3 to 9 years (Table 3.7). Singular subsurface drainage systems, with (pipe or open) field drains directly discharging in an open main drain or natural stream (*Nala*) are even more cost-effective than composite systems. Open drains (up to a depth of 1.0 m) are more cost-effective than pipe drains, but their use is restricted by adverse soil conditions, O&M requirements, loss of land and social settings.

Table 3.7   Economic viability of the composite subsurface drainage systems

	Segwa	Konanki	Uppugunduru	Lakhuwali	Islampur
Net present Value (€ ha^{-1})	1,850	910	1,260	1,270	350
Benefit/cost Ration (-)	1.7	2.8	2.5	3.2	1.2
Internal Rate of Return (%)	58	32	36	39	20
Pay-back period (year)	3	3	3	4	9

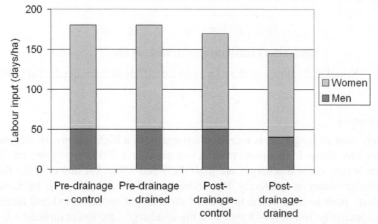

Figure 3.7   Subsurface drainage decreased the workload in rice cultivation – Example from Uppugunduru pilot area.

Farmers' attitude

The attitude of the farmers in the pilot areas, both male and female, towards the new drainage technologies was analyzed by conducting surveys on socio-economic and gender aspects. Almost all famers and their families, total 256, participated in the surveys. The farmers fully realized that subsurface drainage was needed to overcome their waterlogging and salinity problems. Consequently, they contributed to the installation and operation of the drainage systems in various ways. They provided the land for open drains free of charge and organized themselves into (informal) drainage societies. Through these societies, farmers, both male and female, actively participated in the maintenance of the open drains (both field and collector drains) and they took responsibility for watch and ward. Some societies also exerted social pressure on farmers who were initially unwilling to allow collector drains to pass through their field.

Although farmers clearly understood the benefits of subsurface drainage, they are reluctant to pay the total cost, even for O&M (e.g. pumping costs). Like for irrigation, for which they hardly pay any fees, they see it as a government responsibility. On the other hand, after the installation of the subsurface drainage systems, farmers were fully convinced that drainage brings prosperity as, after installation, yields increased considerably. They also realised that their land has improved and problems such as poor land quality, low yields, less quality of fodder and low cropping intensity due to waterlogging and salinity conditions were reduced. They estimated that their asset values increased, on average by 175%.

The survey on gender issues showed that:
- women perception towards the drainage technology had changed positively;
- knowledge on waterlogging and salinity consequences like effects on yield, farm income and standard of living had increased among women;
- the workload of the women in agricultural operations, in particular weeding, was considerably reduced;
- women's' farmers liked the interactions with the project team as it helped them to address some of their farm and home management problems.

Although it is obvious that the farmers, both male and female, are convinced that drainage gives both direct and indirect benefits and prosperity in the long run, they also see the following constraints:
- they prefer the more expensive pipe drains above open drains because with pipe drains they do not have to scarify land and they have to spend less time and money on maintenance;
- they realize that operational and maintenance activities, such as maintenance of pipe and open drains, pumping of drainage effluent, can not be done individually. They realize that these are collective activities for which they have to cooperate among each other;
- they feel that there is a need for government assistance in solving their problems;
- they are willing to contribute, but they expect that the government pays a fair share of the costs;
- they are willing to meet part of the cost or to maintain the field drainage system, but, on the other hand they regard maintenance of the main drainage system as a responsibility of the government.

Farmers clearly see the benefits of drainage. In Gujarat, for example, the economic benefits observed in the pilot areas have already attracted farmers to buy land in the area at prices 5 times higher than during the pre-drainage period, i.e. for € 7,500 to € 12,000 per ha compared to pre-drainage land values of € 1,500 to € 2,500 per ha. Furthermore, farmers from neighbouring areas also started to invest in subsurface drainage. In 2002 and 2003, 15 farmers installed subsurface drainage systems at their own costs: open drains in 45 ha and pipe drains in 3 ha. The project assisted these farmers with the design and supervision during installation. In the other areas, the attitude of the farmers is more reserved, as will be discussed in the next section.

Long-term effects

To verify the recommended drain depth/spacing combinations and to assess the long-term effects, the monitoring programmes continued in two pilot areas, i.e. in Konanki and Uppugunduru. Except for the years 2003 and 2004, when there was no crop cultivated because of water shortages, yields and soil salinity levels show that the performance of the subsurface drainage system is sustainable (Figure 3.8).

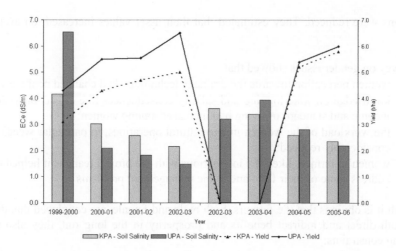

Figure 3.8   Long-term effects of subsurface drainage in Konanki (KPA) and Uppugunduru (UPA) pilot areas: changes in soil salinity and crop yield. Note: in the period 2002 – 2004, no crops were cultivated.

Next, subsurface drainage systems, based on the recommended design criteria, were installed in new areas, i.e. Kalipatnam in the tail end of the Kalipatnam Main Canal in Godavari Western Delta and Mutluru in the middle reach of the Krishna Western Delta [213]. In both areas, soil salinity decreased to acceptable levels after the introduction of subsurface drainage and crops yields increased significantly (Table 3.8).

Conclusions and recommendations

The introduction of irrigation in the arid semi-arid regions of India to sustain agricultural production against the vagaries of the rainfall have resulted in the twin problem of waterlogging and salinity in millions of hectares of good agricultural land. Subsurface drainage, either by open or pipe drains, is an effective tool to combat these problems.

Table 3.8   Effects of the recommended subsurface drainage practices on soil salinity and crop yield in Kalipatnam and Mutluru

Season	Subsurface drainage area		Control area	
	Soil salinity (dS m^{-1})	Yield (t ha^{-1})	Soil salinity (dS m^{-1})	Yield (t ha^{-1})
Pre-drainage:				
Kharif 2004	15.7	3.8	7.0	3.7
Rabi, 2004/05	6.4	5.3	9.4	5.2
Post-drainage:				
Kharif 2005	8.6	3.8	10.3	3.4
Rabi, 2005/06	4.2	8.0	7.2	7.2
Kharif, 2006	--	4.4	--	3.4

Studies conducted in five agro-climatic sub-regions clearly proved that, under the prevailing soils, agro-climatic conditions and social settings, subsurface drainage, by pipe or by open drains, is a technically feasible, cost-effective and socially acceptable technology to reclaim waterlogged and saline lands and to sustain agriculture in irrigation commands. Within one or two seasons after the installation of subsurface drainage systems crop yields increased: on average 54% for sugarcane, 64% for cotton, 69% for rice and 136% for wheat. These yield increases were obtained because in the drained fields water tables and soil salinity levels were respectively 25% and 50% lower than in the non-drained fields. The water tables can be controlled at a relatively shallow depth: 0.50 - 1.50 m (depending on the crop) or even shallower in rice fields. These shallow water tables avoid excessive drainage while at the same time harmful salts that are brought in by the irrigation water are effectively removed. Based on the research findings, the following drain depth/spacing combinations are recommended for the various agro-climatic regions in India:

- *for the medium textured soils in the semi-arid Trans-Gangetic plains of Haryana* (annual rainfall of 500-700 mm): a composite pipe drainage system with a drain spacing of 75 m and a drain depth between 1.10 and 1.50 m. In these light soils a geotextile envelope is required to avoid sedimentation;
- *for the sandy loam to clay loamy soils in the semi-arid east coastal plains of Andhra Pradesh* (annual rainfall of 800 to 900 mm): a pipe drainage system with drain spacing between 60 and 75 m, a drain depth between 1.00 and 1.20 m and a geotextile envelope;
- *for the heavy black soils, mainly cultivated with sugarcane, in the semi-arid plains in Gujarat* (annual rainfall of 1000 to 1500 mm): a subsurface (open or pipe) drainage system with drain spacing of 45 m and drain depth between 0.90 to 1.20 m. In these soils, no envelope is required;
- *for the black soils in Southern Plateau and Hills of Karnataka* (annual rainfall: 770 mm): a subsurface drainage system with open or pipe drains with drain spacing of 50 m and drain depth between 1.00 to 1.20 m. No envelopes are required for soils with a clay content > 50%;
- *for the sandy soils in the arid lands of Rajasthan* (annual rainfall: 300 mm): a composite pipe drainage system with a drain spacing of 150 m and a drain depth of 1.20 m, in combination with a geotextile envelope (non-woven type with a O_{90}-value more than 300 µm).

The recommended drain depths are significantly shallower than the depth traditionally recommended for the prevailing conditions in India (> 1.75 m). These research results support the widely prevailing view that deep drains are unnecessary for salinity control in irrigated lands [234]. The subsurface drainage systems proved to be very cost-effective: cost-benefit ratios are in the range from 1.2 to 3.2, internal rates of return in the range from 20 to 58%, and pay-back periods in the range from 3 to 9 years. Open drains (up to a depth of 1.0 m) are more cost-effective than pipe drains, but their use is restricted by adverse soil conditions, O&M requirements, loss of land and social settings. The use of an envelope is recommended, except for the heavy clay (black-cotton) soils. Non-woven, polypropylene envelope materials performed best, although they are rather expensive. For the construction, a combination of manual and mechanical installation practices, using locally available excavators, was developed and successfully tested. In future, if large-scale implementation is considered, the use of trenchers and trenchless drainage machines, which were already successfully introduced in other projects in India [182; 185; 290] may be considered.

Popular versions of the recommendations, both in English and local languages, were published to disseminate the results to other farmers and local, state and national authorities [202; 214; 240]. These publications also address the challenges to make subsurface drainage a success. First of all, farmers, although they clearly see the benefits of drainage, are too poor to pay the full installation costs. The current government norm of € 300 (about Rs 12,000) per ha for the execution of subsurface drainage projects needs to be revised as it is highly inadequate at the current prices. Next, subsurface drainage can only be successful in controlling salinity, if sufficient good quality irrigation water or monsoon rainfall is available for leaching. Supplementary measures in soil and water management like gypsum application, salt-tolerant varieties, irrigation efficiency improvement, etc. are needed to enhance the positive effects of subsurface drainage. Thus a trade off between the additional investment on soil and water management and saving in cost on drainage should be considered. It should be realized that drainage always has a lower priority compared to irrigation. For the farmers, irrigation is a need for today ('no water no crop') and salinity is a problem of tomorrow. Water users' organizations, not only for drainage but also for irrigation, are not well established. As drainage at farm level, even more than irrigation, is a collective activity, appropriate institutional arrangements for farmers' participation and organization need to be developed. Thus policies have to be reformulated to assure that drainage gets the attention it needs. It is recommended to prepare drainage policy papers at National and State levels emphasising time bound action plans to reclaim waterlogged and salt-affected lands. Only if these challenges are met, India will succeed to protect its capital investments in irrigated agriculture, increase its sustainability and thus be able to feed its growing population.

3.7   Water balance study in a drained area

Irrigation

To investigate the relation between irrigation and drainage, the water and salt balance study conducted in Segwa pilot area, one of the IDNP research sites, is presented. Segwa pilot area is irrigated from various sources: 40% by canal water through the Segwa Minor, 37% by wells, 14% by conjunctive use, 4% by re-using drainage water [225]. Irrigation water supply through the Segwa Minor is of good quality (EC$_i$ = 0.3 dS m^{-1}) and available in abundant amounts. The main crops in the area are sugarcane (43%), rice (21%), fallow (12%) and non-irrigated grassland (25%). The actual canal supplies in the Segwa Minor by far exceed the crop water requirements of sugarcane (Table 3.9).

With the misconception that the more they irrigate, more yields they will get, farmers apply huge quantities of canal water. As the canal irrigation water is heavily subsidized, the withdrawal of groundwater for irrigation is restricted to those parts of the area where there is no canal water supply. The quality of the well water is about 1.7 dS m^{-1}.

Table 3.9   Canal supplies in the Segwa Minor

		Kharif (Jul -Oct)	Rabi (Nov –Jan)	Summer (Feb – Jun)	Year
Actual supply to fields (1)	(mm)	1,332	779	814	2,924
Effective rainfall	(mm)	499	--	--	499
Crop water requirements[a] (2)	(mm)	400	674	1,238	1,912
Excess/Deficit (= (1) – (2))	(mm)	1,332	105	-425	1,012

[a] based on a 65% application efficiency

Surface drainage

In the monsoon-type climatic conditions in India the control of waterlogging and salinity starts with the removal of excess water from the land surface by surface drainage. Timely removal of this excess water is especially important for non-rice crops (Figure 3.9).

The design rate for the surface drainage system is based on the major crop in the area, i.e. sugarcane. Based on a maximum duration of flooding of 7 days and a return period of 5 years, the design discharge rate for sugarcane would be 42 mm d^{-1} (Table 3.10). The effect of adopting this design rate on the other crops has been evaluated. For the same return period (5 years), this design rate results in an additional storage in paddy fields of about 200 mm (Table 3.11). This rise of the level of the standing water is acceptable, except for the first weeks after transplanting, when such a height may damage the crop. This requires controlled surface drainage.

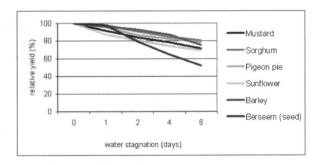

Figure 3.9   Yield reduction caused by water stagnation for various crops (data provided by CSSRI)

Table 3.10   Design rate for surface drainage in fields with sugarcane [161]

7-day rainfall (5 year return period)		658	mm
Storage	0		mm
Infiltration: 2 mm/hr x 24 hr x 7 days	336		mm
E$_o$: 4 mm/days x 7 days	28	+	mm
		364	mm
To be drained in 7 days		294	mm
Design drainage rate (sugarcane)		42	mm d^{-1}

Table 3.11   Additional storage in paddy fields

		Day no.						
		1	2	3	4	5	6	7
Rainfall (P)	(mm)	295	65	118	25	14	78	63
Cumulative rainfall	(mm)	295	360	478	503	517	595	658
Design drainage rate (D)	(mm)	42	42	42	42	42	42	42
Infiltration (I)	(mm)	48	48	48	48	48	48	48
Evaporation (ET)	(mm)	4	4	4	4	4	4	4
Δ storage (= P-D-I-ET)	(mm)	201	172	196	127	47	31	0

Subsurface drainage

Application of the water and salt balance of the root zone using the method described in Section 2.10 indicates that an irrigation efficiency of 80% is more than sufficient to maintain a favourable salt balance (Table 3.12). This requires an average subsurface drain discharge of less than 1 mm d^{-1}. Compared to the actual irrigation application (Table 3.9), it can be concluded that there is ample scope to improve irrigation efficiency.

3.8   Participatory approach

Participatory approaches in subsurface drainage in India are still in a very premature stage of development. Some sporadic attempts have been made, i.e. the example described in the previous section, where in Gujarat farmers, after seeing the effects of subsurface drainage in pilot areas, started to invest in subsurface drainage themselves. Also in the Uppugunduru pilot area in Andhra Pradesh, farmers have been able to manage the subsurface drainage system successfully since July 2003 when they took over the system at the end of the IDNP [213]. The farmers were already used to work together because, being at the tail-end of the irrigation command; they already operated and managed an irrigation pumping station to lift irrigation water to their fields. They created another cooperative society with 36 members to operate and maintain the pumped subsurface drainage system.

Another example of participatory drainage management (PDM) comes from another village in Andhra Pradesh, where farmers read in local news papers over the successful introduction of subsurface drainage in Konanki and Uppugunduru pilot areas. In May 2002, seven farmers of Doppalapudi village in Ponnur mandal of Guntur district approached the IDNP and sought assistance to reclaim their salt-affect lands (about 5 ha) [202]. Their farms had become salinized due to seepage from higher surrounding areas. The situation was so severe that the farmers did not harvest their crops as the rice yields were so low (1.0 to 1.2 t/ha) that it was not economical to do so. The farmers requested technical assistance and expressed their willingness to pay the full cost of installing subsurface drainage. After carrying out pre-drainage investigations, the IDNP team proposed an open drainage system. The cost of construction of the system were estimated to be Rs. 3,600 (€ 63) per ha. With supervision of the IDNP staff, the drainage system was installed by the farmers.

Table 3.12   Water and salt balance in the root zone

Basic Information:

- Land use = Irrigated sugarcane (ratoon crop)
- W_{fc} = 180 mm, no capillary rise
- EC_e = 0.5 EC_{fc}
- EC_e - avg = 2 dS m-1
- EC_e - max = 3 dS m-1
- Eci = 0.8 dS m-1

Step 1: calculate irrigation requirement for leaching   I - year = 1,360 mm (Eq 3)

Step 2: Calculate percolation   R* - year = 181 mm (Eq 4)

Step 3: Select an irrigation efficiency   Irr. Efficiency = 80%

Step 4: Use CROPWAT to calculate irrigation requirement per month

Step 4: Select an irrigation schedule that statisfies the leaching requirements and crop water requirements

Step 5: Change Z1 until dZ = 0

Step 6: Check if ECe < 2 dS m-1 and ECe –year < 2 dS m-1

Period		Year	J	F	M	A	M	J	J	A	S	O	N	D
E	mm	1,868	98	100	199	227	273	186	136	131	144	144	124	107
P	mm	689	0	1	0	0	5	152	176	162	142	28	19	3
E-P	mm	1,179	98	99	199	227	268	34	-40	-31	1	116	104	104
I_{req}	mm	1,262	98	98	199	227	268	36	0	0	11	116	104	104
I_{app}	mm	1,577	122	123	249	284	335	45	0	0	13	146	130	130
EC_a	dS m^{-1}	0.5	0.8	0.8	0.8	0.8	0.8	0.2	0	0	0.1	0.7	0.7	0.8
dW	mm	0	0	0	0	0	0	0	0	0	0	0	0	0
R*	mm	398	24	24	50	57	67	11	40	31	12	29	26	26
Dr	mm	398	24	24	50	57	67	11	40	31	12	29	26	26
Z1	kg ha^{-1}		387	411	432	467	496	519	493	394	331	311	336	359
dZ	kg ha^{-1}	0	24	21	34	30	22	-25	-99	-63	-21	26	23	28
Z2	kg ha^{-1}		411	432	467	496	519	493	394	331	311	336	359	387
ECe	dS m^{-1}	1.1	1.1	1.2	1.2	1.3	1.4	1.4	1.2	1.0	0.9	0.9	1.0	1.0

The saline seepage water from the uplands was intercepted effectively and disposed safely into the natural drain. As a result, crop yields increased to 4.0 to 5.2 t/ha (farmers' estimates). Since then, the farmers maintain the installed drainage system, because they became convinced that drainage is a most appropriate method to reclaim and sustain their salt-affected lands. They realise that the construction and maintenance costs of subsurface drainage are fully justified by the increase in crop yield and that the value of their land has increased tremendously.

3.9   Participatory modelling to cope with off-site externalities of drainage[8]

Introduction

A participatory modelling study was conducted to develop an integrated approach to assess the off-site externalities caused by the disposal of, among others, drainage water in the Kolleru-Upputeru wetland ecosystem on the east coast of Andhra Pradesh. Lake Kolleru (30,855 ha) is the largest freshwater wetland ecosystem in South India [65]. It is located in between the deltaic plains of the Godavari and Krishna Rivers on the east coast of Andhra Pradesh ($16^0 32'$- $16^0 45'$N and $81^0 05'$- $81^0 20'$ E). The Lake is connected to the Bay of Bengal through the Upputeru River or '*Salt Stream*', a 60 km long, intricately meandering tidal river. The Lake is shallow and fresh but, because of its low elevation, brackish conditions prevail in the south-eastern part, especially during dry summer months, due to salt water intrusion through the Upputeru River. The Lake, home for 189 species of birds, including the rare and endangered Grey Pelican, has been designated as a RAMSAR[9] site and was declared a wildlife sanctuary in 1999 [88]. This wetland ecosystem is under threat due to human interventions in the lake itself, in the upper catchment, in the surrounding agricultural lands as well as in the downstream Upputeru River. The main reasons for this degradation are: (i) siltation due to erosion in the upland catchment, (ii) conversion of open water into fish ponds and paddy fields, (iii) pollution with dissolved salts, pesticides and fertilizers from neighbouring agricultural lands, (iv) sewage and industrial waste water from sugar, paper and food processing industries, and (v) salt water intrusion due to reduced outflow and the construction and widening of a straight cut in the mouth of the Upputeru River [27]. This has led to a sharp reduction in the lake's area, the volume of water and excessive growth of weeds and water hyacinth. In combination with over-fishing by the local people, this has resulted in a sharp decline in fish catches, loss of biodiversity, flooding of the adjacent agricultural lands and salinization of the downstream areas. The degradation of the lake not only threatens the fragile ecology but also the livelihoods of the 200,000 local people living in 148 villages and fishing communities in and around the lake [224].

The Government of Andhra Pradesh has recognized the urgent need to stop further degradation and has initiated a number of restoration measures. The latest measure, dismantling of fish ponds below the 1.5 m contour line, has led to fierce opposition from local fishermen and fishpond owners. The opposition became so severe that, at the end of 2006, the government had to enforce a curfew. The main problem is that the restoration measures in general focus on only one of the problems and subsequently often yield less than ideal results. For example, the excavation of the straight-cut (named M29) in the mouth of the Upputeru River to increase the hydraulic gradient and flow rates, resulted in increased salinization during the dry season. Thus, there is an urgent need for an integrated approach.

To initiate this much-needed approach, the Kolleru Lake and Upputeru River ecosystem research project (KLURE) was implemented in 2006. The project was a collaboration

---

[8] Paper 'Participatory modelling to increase stakeholder participation in the restoration of the Kolleru – Upputeru wetland ecosystem in India' submitted for publication in *Environmental Modelling & Software* on 2008-08-08

[9] The Convention on Wetlands, signed in Ramsar, Iran, in 1971, is an intergovernmental treaty which provides the framework for national action and international cooperation for the conservation and wise use of wetlands and their resources (UN Treaty Series No. 14583).

between the Undi Centre of Acharya N.G. Ranga Agricultural University (ANGRAU), Sagi Ramakrishnam Raju (SRKR) Engineering College, Bhimavaram, both from Andhra Pradesh, and Alterra-ILRI, Wageningen University and Research Centre, the Netherlands. A participatory hydrological modelling approach was used to assess the off-site externalities caused by the disposal of drainage water. The drainage effluent not only contains salts and residues of pesticides and fertilizers from the surrounding agricultural lands, but is also heavily polluted with domestic and industrial waste products. The challenge was to come to agreement with the various stakeholders on the outlines of an integrated action plan.

The study is based on IWRM, the recommended approach for developing strategic action plans for the management of lakes in India [188]. In IWRM, stakeholder involvement is seen as crucial [115]. In Europe, this is acknowledged by the EU Water Framework Directive (http://ec.europa.eu/environment/water/water-framework). To make interrelationships between stakeholders explicit and to suggest solutions that are acceptable for all stakeholders simulation models are used. Examples are 'Waterwise' (http://waterwijs.nl), a bio-economic model developed in the Netherlands for spatial planning of lowland basins [271] and 'Aquastress', an EU-integrated project to develop participative modelling tools [135]. A participatory modelling approach that involves local stakeholders with their (tacit) knowledge of the local conditions and circumstances allows researchers to concentrate on the modelling process, rather than the often time-consuming data collection [28]. Participatory modelling can also help to achieve a common understanding or vision how water resource systems function and how they can be managed in a sustainable way [131]. We have adapted these European experiences to increase stakeholder participation in the Kolleru-Upputeru wetland ecosystem.

Materials and methods

A major challenge in using a participatory modelling approach in countries like India is that there is generally a lack of (reliable) data sets, especially of long-term data records. As there was an urgent need to bring the stakeholders together to avoid further escalation of unrest, there was not sufficient time for additional data collection on this hydrological and societal complex ecosystem. Due to this complexity, the decision-making process should be incremental, iterative and continuous [55]. Focusing on dynamics instead of results and focusing on wide-ranging analysis instead of quantitative data are ways to enable progress in conflict-laden negotiations. One of the challenges was to train the researchers and the stakeholders in the correct use of participatory modelling [146]. Based on the IWRM principles and the European experiences with participatory modelling, a four-step approach was adopted (Figure 3.10):

- a reconnaissance survey, based on a literature review in combination with a rapid field appraisal, to collect the additional data needed for the modelling;
- the results of the survey were used to undertake a problem analysis and to prepare data sets for the modelling;
- a simulation model was developed to achieve a better understanding of the complex ecological system, to show this complexity to the various stakeholders and to predict the effects of a number of interventions;
- stakeholder groups were brought together, to give each of them the opportunity to express their (often conflicting) views and interests, to create mutual

understanding that single-issue solutions will not stop further degradation and to find consensus for an integrated approach.

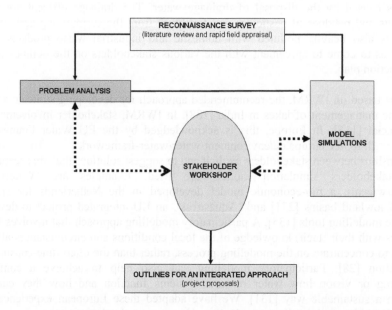

Figure 3.10   Four-step approach for the participatory modelling study adopted by the project.

Reconnaissance survey

Data for detailed problem analysis and input for the model simulations was derived from a literature review in addition to a rapid field appraisal. The following information was collected: topography; land use and land use changes; pollutants entering the ecosystem (both from point and non-point sources); other basin-related causes of impairment like reclamation activities, dredging in the river mouth, road construction, diversion of river distributaries; growth of water hyacinth and other water weeds; ecological values of the wetland ecosystem: flora and fauna and (if available) their interrelations with water quantity and quality; the cultural and legal situation; and existing restoration plans and actions. The latter include measures like dismantling fish ponds, source control, in-lake treatment and shore line management, including people's participation and environmental education and awareness campaigns. The literature review was followed by a rapid field appraisal to collect additional data needed for the model simulations, i.e. cross and longitudinal sections and salinity levels of Upputeru River and Lake Kolleru.

Problem analysis

The results of the reconnaissance survey were used to inform a problem analysis. To achieve a better understanding of its complexity, the Kolleru-Upputeru ecosystem was divided into five components, i.e. the upland catchment, the Krishna Delta, the Godavari Delta, the Lake Kolleru and the Upputeru River (Figure 3.11).

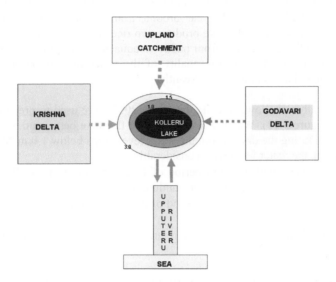

Figure 3.11   Schematisation of the Lake Kolleru and Upputeru River wetland ecosystem.

For each of these components, the main stakeholders, sources of pollution and degradation, and options for restoration were identified using a Strength-Weakness-Opportunities-Threat (SWOT) analysis. As there was a lot of mistrust among the stakeholders, they were already approached during the reconnaissance survey and invited to participate in the problem analysis by bringing in their knowledge and experiences. Integration of the implicit, or tacit, knowledge of the local stakeholder with the explicit, or formal, knowledge held by the researchers was another objective of the project. Combining these two types of knowledge is an essential element of the capacity development process [207].

The Kolleru-Upputeru ecosystem is under threat due to human interventions in the lake, in the upper catchment, in the neighbouring Krishna and Godavari deltas and in the downstream Upputeru River.

The upper catchment (5,400 km^2) accounts for about 80% of the inflow to the lake [241]. The average rainfall in the catchment is about 1,000 mm y^{-1}. Over the past couple of decades, agricultural activities in the catchment have expanded considerably, mainly horticulture and dryland cropping. Because of the intensification in land use, erosion has increased, resulting in high sediment loads in the water flowing into the Lake.
The agricultural lands southwest of the lake, located in the Krishna Delta, do not contribute much to the inflow to the lake. These lands only receive irrigation water from the Krishna River during the monsoon (*Kharif*) season and therefore no water is available for a second crop in the dry (*Rabi*) season. The drainage water that is evacuated to the lake has high salt concentrations (between 1.8 and 8.2 dS m^{-1}) as irrigated agriculture adds about 3.7 t ha^{-1} y^{-1}

of salts to the soil profile, which has to be leached out to sustain the agricultural productivity [203]. Furthermore, the drainage water is heavily polluted with untreated industrial and domestic sewage water [184].

A larger inflow into the lake is received from the Godavari Delta, located northeast of the Lake. Water is available the whole year around from the Godavari River, and the agricultural lands on this side of the lake produce two rice crops per year. Since the 1990s there has been a steadily conversion from paddy to aquaculture on this side of the Lake. This has had a negative effect on the quality of the water discharging into the lake, especially the salt load has increased significantly.

The Lake itself is rather shallow; satellite data from 2002 were used to prepare a contour map of the lake (Figure 3.12). Water levels in the lake fluctuate between 0.5 to 3 m^+MSL (Mean Sea Level). During the dry season, the water level drops below 1.0 m^+MSL. During the monsoon season, the water levels rise considerably [224]:

- upto 1.5 m^+MSL with a return period of 1 year, resulting in the flooding about 15,000 ha of the irrigated lands surrounding the lake;
- upto 2.10 m^+MSL with a return period of 2 years, resulting in the flooding about 33,600 ha;
- upto more than 3 m^+MSL with a return period of 14 years, resulting in the flooding about 57,100 ha.

Siltation has increased considerably due to erosion caused by deforestation and increased agricultural use of the lands in the upstream catchment area. By comparing contour maps prepared in 1967 with the remote sensing data of 2002, sedimentation on the lake bed has been estimated to be around 2.5 cm y^{-1} [26].

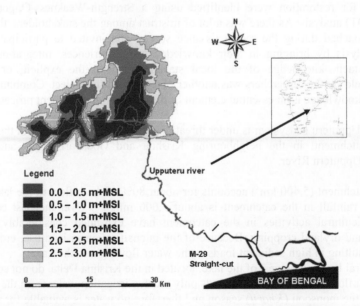

Figure 3.12 GIS contour map of Lake Kolleru and the Upputeru River wetland
          ecosystem.

Table 3.13   Use of Lake Kolleru in February 2001 as interpreted from remote sensing
              data [148]

Land use	Area	
	[km^2]	[%]
Fishponds with water (1050[a])	98.98	40
Fishponds – dried up (38 [a])	4.00	2
Paddy fields	20.97	9
Lake area with dense weeds	57.48	23
Lake area with sparse weeds	53.27	22
Lake area under reclamation	10.30	4
Total	245.00	100

[a]   number of fishponds

Table 3.14   Area of Lake Kolleru converted into aquaculture [183].

Year	Area converted in aquaculture [ha]	Size of the lake [ha]
January 1975	0	16,421
May 1989	116	15,261
May 1995	4,825	11,569
May 1999	6,101	10,320
February 2005	9,191	7,261

Another major change has been reclamation of part of the lake into fish ponds and paddy fields (Table 3.13). Over the past 20 years, this reclamation occurred at an alarming rate, from none in 1975 rising to 9,191 ha in 2005 (Table 3.14). The reduced storage capacity in the lake has increased the risk of flooding in the surrounding agricultural lands (86,000 ha).

The lake drains into the Bay of Bengal through the Upputeru River (Figure 3.12). The average width of the river is about 200 m. Its depth varies considerably with the season but, on average, it is around 3 m. The tidal range in the Bay of Bengal fluctuates between 0.9 and 1.5 m. Because of the low elevation and low discharge during the dry season, sea water intrusion occurs frequently. As a result, the salinity of the water in the southern part of the lake may increase up to 20 dS m^{-1}. To improve the discharge capacity of the Upputeru River the M29 straight cut was excavated in the 1970s. This short-cut shortened the length of the river from 62 to 42 km. To restore ecological sustainability and to improve the livelihood of local fishermen, it is necessary to tackle these multiple problems in an integrated way (Figure 3.13).

Modelling

Simple, easy-to understand models designed in collaboration with the stakeholders are a useful tool to assist in planning [35]. For the model simulations, we used DUFLOW, a one-dimensional, non-steady state, model for water movement and water quality [147], in combination with a GIS-system (Arc GIS 9.1version).

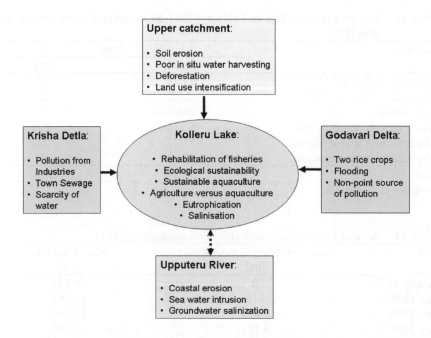

Figure 3.13   Result of the SWOT analysis showing the main problems and challenges.

AutoCAD was used to digitize the topographic maps, to analyze the variation in the lake's area and volume over time and to generate data for the model simulations. One of the major reasons for selecting DUFLOW was that the model was successfully used under similar conditions, i.e. in the Red River Delta in Vietnam and in the Gambia, both complex water management problems with many stakeholders and limited data records [197; 204].

In the Kolleru study, DUFLOW was used to simulate a number of restoration measures, i.e. :
- construction of a weir at the outlet of the Lake Kolleru to regulate the inflow from the lake into the Upputeru River;
- dredging shallow sections in the river;
- closure and re-opening of the M29 straight-cut in the river mouth.

The purpose of these simulations was not to optimize the technical design of these interventions but to increase stakeholder interaction as will be discussed in the following paragraphs.

The upland catchment area and the deltas of the Krishna and the Godavari were not included in the model simulations [241]. The inflow from these areas, both quantity and quality, were used as input parameters. The contour map of Lake Kolleru and twenty-three cross-sections along the Upputeru River were used to model the system. The flow model was calibrated using measured water levels as there were no discharge measurements available. The tidal range and the salinity of the seawater in the Gulf of Bengal were used as a boundary condition. Salinity levels in the river and the lake were also not available, thus salinity calibration could not be performed.

The simulations show that the construction of a regulatory weir (height 3.5 m) will result in higher water levels in the lake and lower water levels in the river downstream of the lake (Figure 3.14). This increase in storage has a positive effect on agriculture as more water becomes available for irrigation and the irrigation season can be extended. A negative consequence is that the water levels in the Upputeru River in the dry season will be lower. Thus, while the weir will decrease the salinity levels in the lake itself, it will increase salt water intrusion in the river downstream of the lake and the risk of salinization in the adjacent agricultural areas.

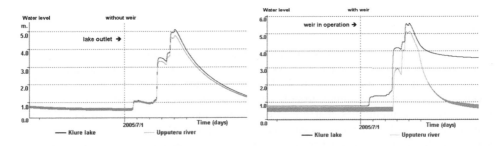

Figure 3.14   Water levels in the Kolleru Lake and Upputeru River: without weir (left) and with a 3.5 m high weir (right).

The effect of opening or closing the M29 straight-cut was also simulated. Since its construction, the M29 straight-cut has been a source of conflicts: local communities that suffered from the increased salt-water intrusion made several attempts to close it again. These attempts were stopped by order of the High Court of Andhra Pradesh (dated 29-03-2004) based on petition made by a voluntary farmers organization (Margadarshi Rythy Club, Apparopet) and the West Godavari Sub-District (the Akividu Mandal). Since that time, the M29 straight-cut is open. Simulations show that about 90% of the annual flow goes through the M29 straight-cut, increasing the salinity levels in the river and reducing the velocity in the old river mouth. This results in a more rapid sedimentation of the old river bed. Finally, the effects of dredging the river mouth were simulated. When shallow sections are dredged, peak discharges in the straight cut will decrease from about 600 to about 280 m^3 s^{-1} and correspondingly increase in the course of the old river (Figure 3.15). The simulations also show that this intervention has not much influence on salinity in the lake, but the increased discharge will reduce sedimentation and salt concentrations in the downstream section of the river.

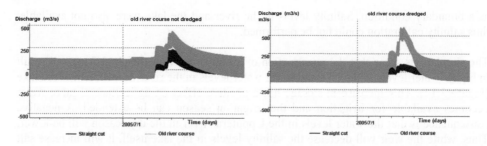

Figure 3.15   Discharges in the M29 straight cut and the old course of the river before (left)
and after (right) dredging the old river mouth to the same level as the M29
straight-cut.

Stakeholder workshop

One of the principles of IWRM is social learning, i.e. learning stakeholders to manage the
issues in which they have a stake [93]. As mentioned previously, there was an urgent need
to bring stakeholders together to avoid further escalation of problems in the area. The
success of participative planning strongly depends on the involvement of those institutions,
businesses and communities that are affected most directly, and interventions that are
appropriate to local circumstances and needs [114]. To achieve these goals, stakeholders
were brought together to identify and explore issues and concerns, to create mutual
understanding that single-issue solutions will not stop further degradation, and to achieve
consensus for an integrated approach. By simulating alternative solutions, stakeholders are
helped to negotiate alternative solutions to stop further degradation [127]. Organizations
and representatives of the various stakeholders were identified during the reconnaissance
survey. These stakeholders were invited to a one-day workshop organized at the campus of
SRKR at Bhimavaram. A reason for selecting this location was that most stakeholders
consider the college as unbiased, so this neutral location was acceptable to most of them.
About 150 stakeholders attended the workshop.

The meeting was chaired by a panel consisting of representatives of the SRKR, ANGRAU,
Alterra-ILRI, an ex-member of Parliament and a member of the legislative assembly. The
stakeholders got the opportunity to present their views on the problems and the most
appropriate interventions they would like to take to solve these problems (Table 3.15).
Their views were used to validate the results of the SWOT analysis (Figure 3.13). Not
surprisingly, there was a big variation in the proposed interventions. After the
presentations of the stakeholders, members of the project team presented the results of their
problem analysis and the modelling process. They highlighted the hydrological processes
in the Kolleru-Upputeru ecosystem and showed the effects of some of the past and
proposed interventions. The focus of the workshop was on the complexity of the problems
and the effects of single-issue interventions. The presentations and discussions clearly
showed that each and every intervention benefits some stakeholders, but also has negative
repercussions to others. In the afternoon session, the problem analysis, the advantages and
disadvantages the proposed interventions and the need for an integrated approach were
discussed.

Table 3.15   Stakeholders' views on the main problems and suggested interventions in the
                  Kolleru-Upputeru ecosystem

Organization/ representative of	Main problems	Suggested interventions
Representative of bed villages of Kolleru	Lack of irrigation water in dry season	Improve irrigation without construction regulatory weir at outlet & compensation for lost land
Farmers' representative of belt villages	Flooding of agricultural lands	Clearing encroachments along the drains and canals and road culverts or even removal of roads
President prawn farmers association	Declining fish catches	Special economic zones for fishery, enforcement legislation and judiciary, alternative fisheries, compensation, waste water treatment
Farmers' representative from downstream part of Upputeru	Salt water intrusion, lack of drinking water	Removal of water hyacinths by farmers themselves
Farmers' representative of Kolleru outlet region	Flooding	Widening Upputeru River and removal of obstructions, no regulatory weir at the outlet
Representative of private land owners in Kolleru Lake	Demolition of agricultural (coconut trees) land, unjustified administrative action against aquaculture activities	Compensation, reconsider the + 5m contour line
Local NGO	Enforcement of Supreme Court's proceedings	Maintenance of drains, no regulatory weir, stakeholders workshops, lake compartment, popular actions to claim land rights
Farmers' representative	Pollution caused by industries and fish ponds	No regulatory weir, compensation
Irrigation Department	Department has no administrative control, drainage maintenance, water hyacinth	Improved methods of weed control
Representative of BJB Minority Morch (NGO)		No regulatory weir, but diversion of Godavari water

Outline for an integrated approach

The main objective of the study was to bring stakeholder groups with conflicting interests together and to achieve agreement on an integrated approach. The main problems were the hydrological and societal complexity of the Kolleru-Upputeru ecosystem, especially the statistical and factual uncertainty on which the proposed interventions are based. To show the stakeholders the complexity of the Kolleru-Upputeru ecosystem, a non-steady state simulation model, DUFLOW, in combination with a GIS system, was used. A literature review, in combination with a rapid field survey to collect additional information, yielded sufficient information for the problem analysis and the input for the model. The implicit (or tacit) knowledge of the stakeholders was linked to the explicit knowledge of the researchers to validate the problem analysis and the model simulations. The main advantage of this approach is that an expensive and time-consuming data collection program could be avoided. This was especially important as quick action was needed as tension in the area was high with frequent conflicts between stakeholder groups.

Combining topographic maps made in 1967 and remote sensing data collected in 2002 proved to be useful to overcome the lack of long-data records. This combined dataset was used to calibrate changes in the lake's topography and associated sedimentation. The model was calibrated using water levels collected during the reconnaissance survey. Simulated and actual water levels in the river were in good agreement. However, simulated water levels in the lake were relatively high, most likely because the lake is quite shallow. After calibration, the model was used to simulate the effects of some proposed interventions. For validation, simulations of an already-implemented intervention, i.e. the closing and (re-)opening of the M29 straight-cut, was matched with the stakeholders' experiences.

The model simulations proved helpful in overcoming potential conflict between stakeholders. Although discussions at the stakeholders' workshop were sometimes fierce, all participants stayed for the whole day and listened to, the often conflicting, views and suggested interventions. The simulations were useful to discuss the effects of these interventions. Simulating past and proposed interventions proved to be a useful tool to create mutual understanding between the various stakeholders that single-issue solutions will not stop further degradation, and formed a basis for creating consensus about the need for an integrated approach. Furthermore, they demonstrated to workshop participants that interventions cannot necessarily satisfy all stakeholders, with beneficiaries and victims associated with each intervention. At the end of the workshop, a mutual understanding for the need for an integrated approach had been reached among the stakeholders. Workshop participants also realized that more research was needed to develop interventions that reverse current negative developments. They expressed the need for technically feasible, economically viable and (especially) socially acceptable interventions. They also stressed that development of these interventions requires political wisdom and common sense and that they should be in accordance with the law.

A follow-up brainstorm session was organized at ANGRAU University. In this 2-day session, a large number of academics, researchers, professionals, representatives of government agencies, NGOs and farmers participated. The outlines of an integrated approach were formulated, addressing the following aspects: sedimentology, ecosystem approach, biodiversity, near-shore ocean dynamics, socio-economic aspects, community participation and hydrology. Subsequently, several more meetings have been conducted and project proposals on the various aspects mentioned above have been submitted by different organizations/institutions to the Central Government. The final approval of the projects is under consideration.

Conclusions

A participative modelling study was conducted to develop consensus for an integrated approach for the restoration of the Kolleru-Upputeru wetland ecosystem. A participatory hydrological modelling approach was used to assess the off-site externalities caused by the disposal of drainage water. The drainage effluent not only contains salts and residues of pesticides and fertilizers from the surrounding agricultural lands, but is also heavily polluted with domestic and industrial waste products. The challenge was to overcome the hydrological and social complexity, i.e. the large variety of hydrological functions and the many stakeholders with different interests. In a four-step approach, the implicit (or tacit) knowledge of the stakeholders was matched with the explicit knowledge brought in by the researchers. Simulating past and proposed interventions proved to be a useful tool to create mutual understanding between the various stakeholders that single-issue solutions will not stop further degradation and to create consensus for an integrated approach. One of the main limitations of using simulation models, i.e. the lack of long-term data records, was tackled by combining data derived from a literature review with a rapid field appraisal and the tacit (or location-specific) knowledge of the stakeholders. During a stakeholder workshop, the outcomes of the resulting problem analysis were matched with the stakeholder's views and experiences. Discussing model simulations with stakeholders proved to be a useful method in creating mutual understanding among the stakeholders about the complexity of the problems. As a result of the project, the stakeholders buried their differences and agreed on the outlines of an integrated approach.

4   Subsurface drainage practices in Pakistan

4.1   History of irrigation and drainage in Pakistan

Agriculture is the backbone of Pakistan's economy, contributing 24% to the gross domestic product (GDP) and providing employment to over 40% of the population [106]. The Indus River irrigation system forms one of the largest contiguous irrigation systems in the world. Like the Nile basin, the Indus basin is one of the oldest and most populated agricultural areas in the world. The country lies in the arid and semi-arid region with an annual evaporation varying between 1,500 to 2,000 mm. The mean annual rainfall ranges from 125 mm in the South-east to 750 mm in the North-West. Rainfall, however, is rather erratic and does not follow the normal monsoon pattern experienced in the region further south. Subsequently, about 80% of the arable land (22 Mha) is irrigated, mainly with water from the Indus River. The Indus River, with a total length of 2,900 km, has five main tributaries with perennial flow, i.e. Jhelum, Chenab, Ravi, Beas and Sutlej. Surface irrigation methods, i.e. basin, furrow, and border are traditionally practices. Agriculture thrives when the rains are on time and properly spaced: then a good cotton crop is followed by a wheat crop. The major crops are rice, wheat, cotton, pulses and sugarcane, besides fruits and vegetables. The average yield of cereals is about 2.2 t ha^{-1}. Before the introduction of the diversion-controlled irrigation in the 19th century a hydrological equilibrium existed between the recharge and discharge of water, enabling a timely removal of excess water and the dissolved salts [73]. The introduction of large-scale irrigation in the 19th century, however, resulted in a distinct rise of the water table (Figure 4.1). As a consequence waterlogging and salinity are now a serious threat to irrigated agriculture: of the 16.7 Mha in the Indus Basin about 1.7 Mha are waterlogged and 2.4 Mha are salt-affected (Table 1.1). The problems were further aggravated because drainage is hampered by construction of roads, railways and/or flood embankments [287].

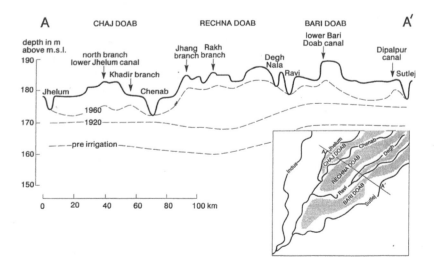

Figure 4.1   Water table profiles in north-eastern Pakistan [43].

To solve the problems of waterlogging and salinity, irrigation canals were lined, supplies were restricted and natural drainage channels that were interrupted by the construction of the irrigation network were restored. These measures, however, were not sufficient to overcome the above mentioned problems and in the 1960s the Government of Pakistan launched a comprehensive plan to control waterlogging and salinity through a series of Salinity Control and Reclamation Projects (SCARP) [73].

The SCARP projects were initiated with loans from the World Bank. The Upper Indus plain was divided into 10 reclamation projects, ranging from 0.4 to 1.6 Mha, and the Lower Indus plan in 16 projects, ranging from 0.3 to 0.8 Mha. Next to the construction of surface drainage systems to restore the natural drainage capacity, 'vertical' drainage was introduced through a network of tubewells with an average density of one tubewell per 250 ha. By the turn of the century, 61 SCARP's were completed, covering about 8.98 Mha. In areas with saline groundwater, the use of tubewells is not very successful because of serious O&M problems. In these areas, mainly located in Sindh, North West Frontier and Punjab provinces, horizontal subsurface drainage systems are being considered more appropriate (Figure 4.2). Subsurface drainage systems have been installed in areas with saline groundwater and cover about 320,000 ha (Table 4.1).

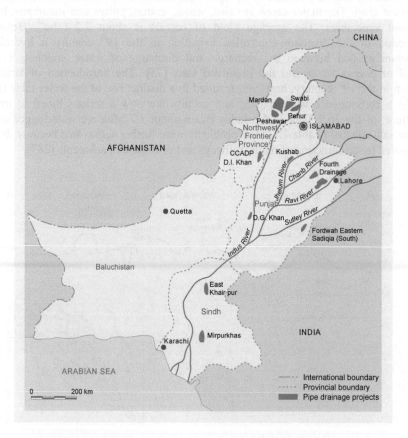

Figure 4.2   Subsurface drainage projects in Pakistan [150]

Table 4.1   Existing drainage facilities in Pakistan [287]

Province	area with surface drainage		area with subsurface drainage		
	GA[a] (Mha)	CCA[b] (Mha)	GA (Mha)	CCA (Mha)	Length (km)
Punjab	4.394	3.888	0.095	0.082	2,810
Sindh	2.726	2.313	0.018	0.024	2,046
NWFP	0.358	0.294	0.266	0.213	7,756
Balochistan	0.072	0.065	-	-	-
Total	7.550	6.560	0.380	0.320	12,612

[a] GA = gross area   [b] CCA = cultivatable command area

4.2   Organization of the drainage sector

Water and Power Development Authority

In 1958, the Water and Power Development Authority (WAPDA) was established as the agency responsible for the coordination of design, construction and initial operation of the engineering works [150]. WAPDA is responsible for the design, construction and initial operation and monitoring of the SCARP projects, after which the Provincial Irrigation Departments (PID) take over O&M.

Area Water Boards and Farmer Organizations

As the drainage fees cover only around 20% of the actual expenses of O&M, the financial burden to operate and maintain the public tubewell systems became gradually too much for the PIDs. These problems were aggravated because the life expectancy of most SCARPs proved to be less than half the expected life time. To overcome these problems, the irrigation and drainage sector was reformed and in 1997 Provincial Irrigation and Drainage Authorities (PIDA) were established in all four provinces [37]. System management is to be decentralized and farmers are to take part in the system development and to take over O&M. This is realized by the creation of Area Water Boards (AWB) and Farmer Organizations (FO). PIDAs facilitate and promote the formation of AWBs, which are composed of farmers, government and PIDA representatives. AWBs on their turn facilitate and promote the formation of FOs. The PIDAs are responsible for the planning, construction, O&M of the system at main and secondary level. At tertiary level, the FOs are responsible for O&M of the system. All these organizations have to become financially autonomous by levying water charges and drainage fees. The establishment of FOs and AWBs is however hampered by (i) a lack of farmers' involvement in policy reforms, (ii) the weak legal framework (the PIDA Acts) to implement reforms, (iii) lack of knowledge within the FOs and AWBs to develop and implement strategies to deal with the systems' problems and (iv) reluctance of the authorities to make the shift from engineering to institutional solutions.

International Waterlogging and Salinity Research Institute

In 1986, the International Waterlogging and Salinity Research Institute (IWASRI) was established. IWASRI, which is part of the WAPDA, has the mandate to conduct, sponsor, manage and undertake research on waterlogging and salinity in Pakistan. In 1988, the Netherlands Research Assistance Project (NRAP) was initiated, a joint undertaking by the International Waterlogging and Salinity Research Institute (IWASRI), Lahore, Pakistan and the International Institute for Land Reclamation and Improvement (ILRI), Wageningen, the Netherlands [21]. The project, which covered the period 1988-2000, had two main activities: work on technical aspects of drainage and the development of a participatory approach to drainage.

4.3   Need for subsurface drainage

Subsurface drainage systems are rather expensive, fortunately there are several options to reduce the need and/or intensity of these systems, e.g.:
- improving irrigation practices to reduce subsurface drainage requirements;
- improving surface drainage to reduce subsurface drainage requirements;
- interceptor drains to minimize drainage requirement;
- lining of irrigation canals to reduce seepage;
- groundwater modelling to identify areas in need of subsurface drainage.

Improving irrigation practices to reduce subsurface drainage requirements
Pakistan is a water deficit country. A typical water supply of its large-scale, low-supply, irrigation schemes would be 3.5 cusecs/1,000 acres, which equals 2 mm d^{-1}. This supply is by far not enough to satisfy the crop water requirements. The systems are designed as *'protective irrigation'*, based on proportional division of water over the available land. A great improvement in productivity is expected of better matching irrigation supplies with crop demand. There are several attempts to include this match in projects, e.g. on the Fordwah Eastern Sadiqia South (FESS) project. In FESS an attempt was made to get a closer match of water deliveries based on the crop water requirements by improved scheduling of delivery of water through the introduction of structural, operational and management improvements. IWASRI/NRAP analysed, together with IIMI-Pakistan, the possibilities to introduce irrigation based cropping [293]. The availability of water in Pakistan, however, is not sufficient for crop-demand based supply of canal irrigation water, with the capacity of the existing reservoirs fully utilized. Hence, a shift to crop-based supply in one scheme cannot be done without affecting the water share of other schemes. Moreover, the capacity of the canal system in Pakistan is not sufficient for crop-demand based supply of irrigation water. Matching crop requirements would also result in demands that vary considerably over time. This would require another system, with much more regulation flexibility, and a more intensive operation throughout the seasons. Moreover, the sediment load of the water prevents canals to run at less than 70-75% of their design capacity. It appears that efforts towards crop-demand based supply end up in recommendations towards irrigation based cropping. In a water deficient situation, moving towards demand-based operations is beset with problems. It will be better to improve the performance of the present water allocation than to respond to field-generated demand that cannot be satisfied. The possibility to achieve a better match between crop water requirement and delivery of water through introducing structural, operational and management improvements is very limited. In many canal systems it seems better to just

keep the supply constant and let the farmers pump from tubewells to complement the shortage of canal water.

Improving surface drainage to reduce cost of subsurface drainage

Contrary to irrigation canal systems, surface drains are not self-cleaning, restricting the existing capacity of these drains. In two projects, i.e. FESS and Fourth Drainage Project (FDP), one of the first actions was the improvement of the surface drainage systems [112; 120]. Drains were desilted and, at some locations, inlets were constructed to guide possible overland flow. Monitoring programmes clearly indicate that after the surface drainage was improved, the water tables are significantly lower (Figure 4.3). It can be concluded that investments in maintenance of surface drainage systems, which are generally low compared to investments in the implementation of subsurface drainage, can significantly reduce the need and/or capacity of subsurface drainage.

Interceptor drains to minimize drainage requirements

Seepage from irrigation canals has long been considered one of the major contributors to waterlogging and salinity. Consequently, the idea was that when this seepage could be intercepted with interceptor drains along the canals, the need for drainage would be considerably reduced. Therefore, implementation of interceptor drains was proposed in various areas in Pakistan.

Lining of irrigation canals to reduce seepage

An option to reduce seepage is lining. Research conduced in Pakistan shows that lining the irrigation canals can reduce seepage significantly [18]. The impact these seepage rates have on subsurface drainage systems is, however, negligible: only about 0.06 mm d^{-1} on a design discharge of 1.5 mm d^{-1}. Furthermore, the seepage is often only a few percents of the canal flow.

Groundwater modelling to identify areas in need for subsurface drainage

In large-scale irrigation projects, subsurface drainage is in general installed in the total project area. Analysis of water table data and inverse modelling with the groundwater model SGMP in the FDP project showed that subsurface drainage was only needed in about 60% of the project area. Furthermore, the design discharge rate could be reduced based on the variation in the natural conditions and the location and capacity of the existing water courses [39].

4.4   Design principles

Layout

The subsurface drainage systems installed in Pakistan typically consist of a composite system consisting of a buried collector and field drains. The major parts of the irrigated areas in Pakistan where waterlogging and salinity occur have little slope, basically 'one foot per mile' (= 0.20 m km^{-1}), therefore pumped subsurface drainage systems are required and most collectors discharge into a sump from which the water is pumped into an open drainage network.

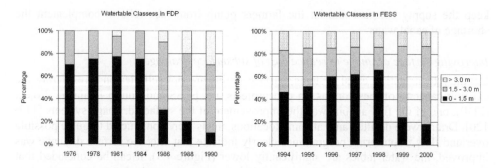

Figure 4.3   After the surface drainage system was rehabilitated, water tables were significantly lower - left: FDP rehabilitated in 1985 [120] and right: FESS rehabilitated in 1999 [112].

Design criteria

The drainage design discharge is a function of crop, water and leaching requirements and varies between 0.95 and 3.5 mm d^{-1} (Table 4.2). Drain depths are relatively deep, basically because of two reasons: (i) to reduce salinization of the root zone through capillary rise and (ii) deeper systems were cheaper because deeper drains allow larger drain spacing.

Drainage materials

Corrugated perforated PVC pipes are used for field drains (Table 4.3). As the hydraulic conductivity of the soil is relatively high (e.g. in Mardan Scarp in the Northwest Frontier Province the hydraulic conductivity ranges between 1 and 3 m/d), the spacing between field drains is large and consequently the pipe diameters are rather big (100 < Ø < 200 mm). Collectors are also made of perforated PVC pipes, with diameters between 200 and 380 mm.

Table 4.2   Drainage design criteria for some major projects in Pakistan [38].

Project	Designed	Design parameters			
		Discharge (mm d^{-1})	Drain depth (m)	Depth of water table (m)	Hydraulic head (m)
East Khaipur Tile Drainage	1976	2.5 – 3.5	1.95	1.0	0.95
Mardan SCARP	1983	3.0	2.40	1.05	1.20
Fourth Drainage	1983	2.44	2.40	1.20	1.20
Chashma CAD	1984	1.2 – 2.6	2.10	1.40	0.90
Fordwah Eastern Sadiqia (South)	1994	1.5	2.10	1.20	0.90
Khushab SCARP	1990	1.8	2.10	1.20	0.90
Swabi SCARP	1994	2.0	1.80	1.00	0.80
Mirpurkhas II	1994	0.95	1.80 – 2.40		
DC Khan SCARP	1995	1.88			

Table 4.3   Materials used in some major drainage schemes in Pakistan[150] [150]

Project	EKTDP[a]	FDP[b]	Mardan[c]	Khusab
Size of the unit	280 - 450 ha	380 - 400 ha	100 - 300 ha	
Field drains	100 mm corrugated PVC pipe	100 - 200 mm PVC pipe	100 mm PVC pipe	100 & 150 mm PVC pipe
Envelope	Gravel	Natural river run gravel with hydraulic conductivity >15 m d^{-1}	Mainly gravel but for some collectors synthetic fabric envelope was used	Gravel mainly but for IWASRI research synthetic envelope was used on one sump.
Collector Drains	225 - 450 mm CC and 250 - 300 mm PE pipes	Perforated 250 - 375 mm PVC pipes	Perforated 100, 150, 188, 250 & 300 mm PVC pipe	100, 150, 200, 250, 300 & 380 mm PVC pipe
Sumps	Pumps provided	79 circular sumps, with 1 to 3 pumps of each 0.06 m s^{-1} capacity	Outflow was by gravity and therefore no sump required	56 sumps: 45 circular and 11 rectangular

[a]   East Khairpur Tile Drainage Scheme, Sindh Province
[b]   Fourth Drainage Project, Faisalabad, Punjab Province
[c]   Mardan SCRAP, Northwest Frontier Province

Most soils in Pakistan are fine-textured (silty loam, sandy loam, silty clay, etc.) and thus require an envelope. In general, gravel envelopes are installed using design criteria developed in the USA [281]. In several projects, problems with the use of gravel envelopes were encountered and some of the following improvements were successfully introduced [150].

The trench box of the drainage machines was modified, because it was observed that gravel was not laid uniformly around the pipe. The trencher box was equipped with an auger that rotates around the pipe below the gravel box feeder. The speed of this gravel auger is automatically adjusted to the speed of the trencher during drain installation. This modification was first introduced in FDP and subsequently improved in the CCAD and the FESS projects. The results are encouraging and gravel is laid comparatively uniformly. In the CCAD project, the supply of gravel under the wet conditions was problematic: although the trencher, equipped with its wide tracks, performed satisfactorily, the performance of the auxiliary equipment like gravel trailers and excavators was poor.

Serious problems occurred with the crushed rock envelope used in the FDP although it was designed according to the specifications [280]. The design specifications, which were based on the United Stated Bureau of Reclamation (USBR) criteria, specified that well-graded gravel with a minimum thickness of 100 mm should be placed around all pipe drains. Normally river-run gravel is used in Pakistan, but because river-run gravel was not available in the vicinity of the FDP area, the use of crushed gravel was proposed by the contractor. Soon after installation started it became clear that the drain lines for which the crushed gravel was used did not perform satisfactorily: drain pipes were chocked by soil that had entered the pipe. The execution was stopped to investigate the cause of the problem. Drains were excavated and it was discovered that a lot of fine soil had moved into the drains. Subsequent laboratory tests revealed that the hydraulic conductivity of the crushed gravel (> 900 m/d) was much higher than river-run gravel of the same gradation (75 – 250 m/d) [221]. It was concluded that the resulting higher hydraulic gradient had allowed the finer soil particles to enter the pipe.

A gravel envelope is also rather expensive. In the EKTD project, for example, the cost of the gravel envelope material, including transport (€ 205/ha) was 17% of the total cost of installing the subsurface drainage system (€ 1,183/ha), about the same as the cost of the pipe material (€ 236/ha) and double the cost of the installation (€ 100/ha).

4.5   Installation practices

Drainage in Pakistan is generally executed within the canal irrigation commands. The drainage projects are contracted to public or private consortia under the authority of WAPDA. Both chain-type trenchers and trenchless ploughs are used for the installation. The trenchless ploughs are not very efficient: due to the traction conditions, towing services of an additional crawler tractor (225 KW) were normally required.

As was mentioned in Chapter 1, sound engineering judgement on the spot is a prerequisite for success, some of the lessons learned with subsurface drainage practices in Pakistan are summarized below.

Planning
In the Mardan SCARP project, discontinuation of irrigation a few days before and during installation is required to obtain sufficient grip for the drainage machines [251]. In an area like this, which is intensively cropped and has many (small) farm holdings, a good coordination between the landowners, farmers, contractor and engineer is essential for a smooth work process. Frequent and jointly organized inspections are essential to ensure good quality installation practices and specifications of construction requirements, inspection procedures, etc. have to fully and carefully define the requirements of the works. They must also address any unique problems that are likely to be encountered during the work. Again these specifications need to be developed in close cooperation between the consultant and the contractor.

In the CCAD Project, a feasibility study was not conducted and the project was commissioned based on the limited available information [150]. Investigations, surveys and designs were only carried out after the project execution started. This resulted in many changes of the original plans. Although this delayed the project for several months, millions of rupees were saved that would otherwise have been wasted on unnecessary drains if the project had been executed according to its original design. The equipment used in this project suffered excessive wear and tear due to the extremely wet conditions. The digging chains and allied parts of the trencher machine wore very rapidly due to the abrasive action of sand. Replacement of these digging chains in the CCAD project was eight times more than for similar projects in Pakistan: after digging 3.5 – 4 km of trench in the CCAD project area compared to e.g. 30 km of trench in Nawabshah. Another reason for this was the contractor's procurement of locally manufactured chains. Replacement of a digging chain costs 2 working days.

Type of materials
In the EKTD project, in the Sindh Province, the installation of the concrete collector drain pipes was a cumbersome and costly job [150]. Prior to the installation of the collector pipes sections of the collector line had to be dewatered by horizontal dewatering and some sections even by vertical well-pointing due to the unstable soil conditions in the area. Soon after installation it became clear that the performance of the concrete collector drain pipes was unsatisfactory. The unstable subsoil caused dislocation of the concrete pipes, sink holes appeared, and costly repairs were necessary. So, it was decided to install large diameter perforated PE pipes with a gravel envelope in the remaining collector units. The PE pipes had to be imported, as large diameter PE or PVC pipes were not yet locally made. The installation and performance of the PE collector drain pipes proved to be successful in unstable soil. So, in unstable subsoil no concrete drain pipes are to be used but only perforated collector drain pipes with envelope material.

In the FDP, the main lesson learned from this project is that specifications based on knowledge that was developed elsewhere (in this case the USBR criteria, see Section 4.3.4) will have to be locally verified during the project's preparation phase [179].

Trench backfill
In Pakistan, the soils in the areas in need for subsurface drainage are relatively fine-textured soils. Consequently, drain spacings are wide and thus field drains and collectors are deep, sometimes up to 4 m near the sumps. In several projects, sink holes appeared after the installation of drains [150]. The reasons were that, although the consolidation of the top layer was reasonably good after backfill, the conditions immediately above the drain pipe were poor and did not improve in time. This was because consolidation of the backfill on top of the drain pipe in semi-saturated conditions was not possible, as the equipment used for compaction could not go deeper than 1.5 m. Just after installation, the trench often collapsed resulting in large humps of soil on top of the drain pipe resulting in big voids. The sink holes appeared as the result of piping after irrigation and rainfall events. Sometimes, sink holes appeared only two to three years after construction especially when the trench backfill had not been exposed to irrigation and/or heavy rainfall events that are needed to consolidate the trench. To reduce the risks of sink holes, excessive gradients were avoided by reducing pumping from the sumps during construction. Pumping was resumed only after trench backfill has been exposed to one cropping season irrigation and/or to a heavy rainfall event. Furthermore, additional measures like rollers, puddling, extra soil, blinding, slow water table drawdown and deep tillage were used to overcome this problem.

4.6 Participatory drainage development

The farming community is generally not involved in large-scale projects. During the last three decades, several small-scale subsurface drainage systems were designed and constructed on famers' land on a cost-sharing basis. In FESS, a subsurface drainage system was installed in a 112 ha pilot area with the assistance of an NGO, Action Aid Pakistan, and in close cooperation with the farmers [21]. The area was selected based on a topographic survey and a participatory rural appraisal. Meetings with the farmers and the involved government agencies were organized to agree on the farmers' contributions. Farmers agreed to (i) assist with data collection, (ii) provide unskilled and semi-skilled labour, (iii) cash payments and (iv) organize work and tasks. Farmers were involved in designing the system: they had a major say in selecting the location of the sump and the layout of the field drains was adjusted so that more farmers could benefit.

Initially it was agreed that the drains should be installed manually, but high water tables made this impossible and consequently the subsurface drainage system was installed mechanically. Farmers, however, dug 'dewatering' trenches along the drain line to prepare the top soil for the weight of the drainage machines. During the actual project implementation some farmers were more motivated than others, their interest depended on total land holding, extent of the waterlogging and salinity problem, farmers' dependency on agriculture, conflicts between farmers, lack of leadership, etc.

A Farmers' Drainage Organization, established in 1997, gradually took over the responsibility for O&M. A gender programme was included to emphasize the role of women, mainly as motivators of their men to participate in and contribute to the project. During the implementation of the project, training courses were organized for the farmers, project staff and staff of the NGOs (Section 5.1.6). These training courses were highly practical and designed to transfer information between the stakeholders with the overall aim to make the operation of the drainage system easier. The cost of the system was about € 526/ha (2000 prices) of which the farmers contributed about 10%.

Similar systems were constructed in the Lower Indus Basin (Table 4.4). A 10-year monitoring programme showed that these participatory drainage schemes effectively control waterlogging and salinity which made it possible for most of the famers to recover their capital costs [118]. The farmers' contributions indicate that they are willing to contribute. In these projects, much time and effort (examples) was needed to build confidence among the farming community. It was recommended that tax and duty exemption, along with interest-free loans, would be provided to assist the farmers to install on-farm drainage at their own expense.

These and other experiences show that tertiary-level drainage beneficiary groups may be effective for supporting project financing and implementation but that they are less relevant for management [296]. Because of this and because of economics of scale, it is recommended that these type of drainage beneficiary groups should become part of farmers' organizations operating at the secondary canal command level (about 3,000 to 10,000 ha).

Table 4.4   Participatory drainage development, some examples from the upper and lower Indus Basin [21; 118]

Scheme	Year of Construction	Area (ha)	Farmers' share (%)
Upper Indus Basin:			
Bahawalnagar, Punjab	1998	112	10
Lower Indus Basin:			
NIA farm, Tando Jan	1987	4	25
Bughio farm, Mirpurkhas	1988	17	30
Nawazabad farm, Hyderabad	1989	41	60

4.7   Operation and maintenance

Formally, O&M of drainage systems is to be taken care of by the Provincial Irrigation Departments, a few years after completion of the systems. However, these Departments do not receive additional funds and therefore, the systems could not be operated and maintained as necessary. Beside the lack of funding there are other problems such as power failure, mechanical problems and lack of farmers' cooperation. Due to this very often the drainage benefits expected at the time of design cannot fully be achieved. IWASRI has reviewed the performance of drainage systems to assess the problems with O&M and the possibilities to increase farmer's involvement. It was concluded that not too much can be expected of this type of '*social approach*' in a short time, because [21; 123-125; 180]:

- farmers might be ready to pump for irrigation, but they will not pump 'continuously' for drainage;
- the resource base of the small farmers is very narrow. Small farmers cultivate about 45% of the land in Pakistan. They typically have a farm size of less than 2 ha and they have virtually no own resources. Moreover, the price they can get for their products is often well below the market price, which in Pakistan, is even lower than the international market, or they have to pay water fees even when they don't receive canal water;
- sincere involvement of farmers takes time. Several current, hurried, attempts to promote participatory approaches in on-farm drainage stand little chance of success. Even with a properly functioning main drainage system, and a favourable attitude of users and bureaucracy, it would be time-consuming, and;
- there seems to be, at decision-taking level, a lack of understanding of what it takes to involve farmers, especially in the planning, implementation, and O&M of drainage systems.

4.8   Disposal of the drainage effluent

In Pakistan, about 9 million tons of salts, dissolved in the drainage water, are discharged annually into the Indus River causing major water quality and environmental problems [6].

Safe disposal of the drainage effluent is complicated because the majority of the agricultural lands, about 10.0 Mha of the total 16.7 Mha, are located in the Punjab in the upper reach of the Indus River Basin (Figure 4.2). The government encourages reuse of drainage water for irrigation in conjunction with the canal water supplies [33]. Not all drainage effluent from the agricultural lands in the Punjab, with a salinity that can vary between 4.7 and 15 dS m^{-1}, can be reused nor discharged back into the river system: the downstream salinity becomes too high. Two alternative options to dispose the drainage effluent are implemented: (i) outfall drains that bypass the Indus River and discharge directly in the Arabian Sea and (ii) evaporation ponds.

Left Bank Outfall Drain

To create a safe outlet, the Left Bank Outfall Drain (LBOD) was constructed to convey the highly saline subsurface drainage water from an area of about 577,000 ha directly into the Arabian Sea [33; 73; 139]. The LBO drain is 250 km long and has a capacity at the outfall of 113 m^3 s^{-1}. For the Drainage Master Plan of Pakistan [287] the need and design for this outfall drain was assessed using the DRAINFARME approach [230]. A pilot survey was carried out for the Kotri sub-basin that includes the entire left bank delta of the Indus River. The outcome suggests that for many years to come drainage problems can be solved at basin level. To achieve this, however, the water management, including drainage, has to be improved and the institutional weakness addressed.

Evaporation ponds

If there is no safe outlet for the drainage effluent, evaporation ponds can be used. Evaporation ponds have been constructed to dispose drainage water from irrigated areas bordering the desert towards the Southeast of the country [33]. These areas are located 500 – 800 km from the sea and they are characterized by interdunal depressions with highly sodic soils lying between longitudinal sand dunes 4 – 9 m high. In the ponds, the drainage effluent evaporates from the open water surface, leaving the salts and other soluble trace elements behind [196]. Attempts to dispose the drainage effluent in evaporation ponds have not been very successful because evaporation ponds need quite a large area, between 10 and 15% of the land, and because of environmental constraints, i.e. seepage of saline drainage water, both from the unlined drains as well as from the evaporation ponds itself, pollutes the groundwater reservoirs. IWASRI conducted a field study to investigate the effects of evaporations ponds that were developed to dispose off the drainage effluent of 514 drainage tubewells, installed to alleviate waterlogging in the SCARP VI area. The evaporation ponds consists of a series of inter-dune depressions locally known as 'Tobas'. A literature review in combination with a field study was conducted to assess the environmental impacts of these ponds. The main findings are not very encouraging [36]:

- the water balance indicates that ponds are not very effective: on average only 12% of the incoming water evaporates. One of the reasons is that the evaporation from an brackish evaporation pond is about 15% lower than from a fresh-water pond [113];
- to increase the effect of these evaporation ponds, the inflow has to be reduced significantly, for example by changing the drainage technology from tubewells to subsurface pipe drainage: not only the quantity of effluent can be reduced significantly but also ponds will be more sustainable due to reduced salinity levels;

- the saline water can be used for saline agro-forestry, fisheries and salt mining. Eucalyptus is the best tree species for agro-forestry and most of the farmers are willing to grow these trees. The present salinity of the pond water is suitable for fish species like *Tilapia Mosambiqa* and *Tilapia Nolitica*. However, the food required for prawn culture in existing saline ponds is not available. There is potential for salt mining as no heavy metals were found in the water. This mining, however, depends on the operation of saline drainage tubewells and is limited to 3 million tons of sodium chloride per year;
- lateral seepage from evaporation ponds badly affected the adjacent agricultural lands. A lot of agricultural land around the ponds became waterlogged and has gone out of production. This badly affected the life of the people living in the vicinity of ponds: 2 out of 12 affected villages were completely abandoned and of the other villages 49% of the residents moved out;
- environmental degradation can be balanced through agro-forestry in areas affected by salinity and waterlogging. A combined approach of surface, subsurface and bio-drainage is most suitable for the problems of waterlogging and salinity in the project area.

5   Improving subsurface drainage practices

5.1   From manual installation to large-scale implementation[10]

5.1.1   Introduction

Subsurface drainage is a form of drainage that was widely introduced in Europe and North America in the 20th century. Egypt is the country with the largest area provided with subsurface drainage, about 2.5 Mha [24], while countries such as Pakistan, China, Turkey, and India are providing subsurface drainage to large tracts of their irrigated lands [79; 105]. Subsurface drainage has a long history, the oldest known systems date back from 3000 B.C in Mesopotamia. Drain pipes were already in use some 4000 years ago in the lower Indus Valley and bamboo pipes were used as drains in ancient times in China [41]. Pipe drainage in modern times started in the United Kingdom in the 17th Century in the form of trenches filled with bushes or stones and was further developed on a large scale in the Netherlands [261]. The first clay pipes were produced in 1810, followed by concrete pipes a few decades later. The necessary envelop material around the field drains originally consisted of locally available materials like stones, gravel or straw.

In the first half of the 20th century, the prevailing empirical knowledge of drainage and salinity control gained a solid theoretical footing [193; 229; 290]. This enabled the introduction of subsurface drainage in many parts of the world. This introduction was further accelerated by the rapid developments in mechanized installation from the 1940s onwards [163; 217]. That the rapid development did not come without problems is well illustrated by two quotations from Van Schilfgaarde. In 1957 he wrote: '*Notwithstanding the great progress of recent years in the development of drainage theory, there still exists a pressing need for a more adequate analytical solution to some of the most common problems confronting the design engineer*' [269]. It was, however, not so much the lack of a theoretical background that hampered the introduction of subsurface drainage, but rather the practical tools needed for the implementation. In 1978, the same author summarized the state of the art for the International Drainage Workshop at Wageningen as: '*Not much will be gained from the further refinement of existing drainage theory or from the development of new solutions to abstractly posed problems. The challenge ahead is to imaginatively apply the existing catalogue of tricks to the development of design procedures that are convenient and readily adapted by practising engineers*' [270].

To meet these challenges, a number of problems had to be solved. Firstly, the traditional use of clay or concrete for drain pipes and organic materials or gravel for drain envelopes resulted in: (i) high transportation and installation costs because of the weight and shape of the materials, (ii) poor quality of construction because of the large number of pipes involved, with each and every pipe-joint creating a weak point in the system, and (iii) the rapid decay under semi-arid and arid climatic conditions of the organic envelopes traditionally used in Europe. Secondly, the traditional method of quality control, i.e. post-construction, e.g. checking the grade of the drain pipe after installation, proved to be

---

[10] Published as: Ritzema, H.P., Nijland, H.J., Croon, F.W., 2006. Subsurface Drainage Practices: From Manual Installation to Large-Scale Implementation. *Agricultural Water Management*, **86**, 60-71

inadequate because of the increased speed and method of installation. Thirdly and finally, the introduction of modernized drainage machinery and installation techniques demanded experienced engineers, operators, technicians and foremen, as well as proper planning and organization of the implementation process. Innovative training was needed to fulfil these demands. In this section, I address how these challenges have been met.

5.1.2 Installation equipment

Excavating and trenching machines, driven by steam engines, were introduced in 1890, followed in 1906 by the dragline in the U.S.A. [163]. The invention of the fuel engine in the 20th century has led to the development of new machines. First the so-called trencher machines were introduced, followed in the late 1960s by the introduction of trenchless machines [44; 50; 51; 150].

Installation in trenches

Trenchers dig a trench at the required depth and grade and place the drain pipe at the bottom of the trench. Several types of trenchers are produced in various sizes and a wide range of capacities. They can install pipes to a depth of about 3 m in trenches up to 0.50 – 0.60 m in width. A trencher can install all known types of pipes and envelopes. They have a maximum speed of installation of 2 km per hour and an average output of 1.5 – 2.5 km per day, depending on the logistics supporting the machine (Table 5.1). Trenchers have been modified in various ways so that they can also be used to install drains in stony soils, in orchards, or in soils with high water tables [150].

Trenchless installation

Machines for trenchless installation do not excavate a trench but act as a plough, with the soil being lifted while the pipe is installed. Two types of machines are used: subsoilers with a vertical plough and V-ploughs. Installation speed is higher compared to the trencher machines, i.e. maximum of 4 km per hour and an average output of 4-5 km per day (Table 5.1). The average output, however, is low because on the non-working time, i.e. time that the field conditions do not allow installation, time for repairs, time for regular maintenance, time for meals, time for non-working days, i.e. weekends, holidays and crop stoppage periods, and time for organisation, i.e. driving to the correct location [142]. Trenchless machines have more restrictions than trenchers: the maximum installation depth is about 1.8 m and only corrugated plastic pipes and pre-wrapped envelopes can be used. The advantages of trenchless drainage decrease rapidly with greater drain depth and heavier soils (Table 5.2).

Capacities, support equipment and further requirements

Both types of machines have been introduced in many places in the world. Their capacities and corresponding installation costs vary from place to place, depending on the local physical and economic conditions (Table 5.1). Generally speaking, the cost of large-scale pipe drainage systems ranges from € 750 to € 1,500 per ha.

Table 5.1   Installation capacities, machine cost and total cost of large-scale subsurface drainage projects in various countries.

Installation method	Country	Capacity [a] (m hr⁻¹)	Machine cost [b] (€ km⁻¹)	Total cost [b] (€ ha⁻¹)	Reference
Trencher:					
Collectors [c]	Egypt	55-100			[149]
Field drains [c]	Egypt	190-380	840	400	[149]
Field & collector drains	India		1,820	770	[150]
Collectors	India	300		778	CADA [c]
Field drains	Netherlands	400	340		[149; 154]
Field drains	Pakistan		950	1,183	[154]
Field drains	USA	100-200	700-800	1,025	[44; 154; 163]
Trenchless:					
Field drains – V-plough	Egypt	626			[63]
Field drains – Vertical plough	India	500-750		778	[150]; CADA [c]
Field drains – V-plough	Netherlands	430-1,150			[272]
Field drains - Plough	USA	400-1,250	500-1,200	1,025	[44; 163]
Combined mechanical/manual installation:					
Excavator	India	187.5			[150]
Backhoe	India	62.5			[150]
Excavator	India	2		385	[103]
Tractor	Netherlands	188.5			[149]
Collector	Pakistan		7,050	1,183	[154]
Backhoe	USA	12-24			[44]
Manual installation:					
Field drains	India	0.25		495	[103]
Collectors	Netherlands	5.5			[150]

[a] effective hours;
[b] cost figures are only indications: actual costs vary considerably due to exchanges rates, year of construction, etc.;
[c] personnel communication, 1999

To further increase the efficiency of both trenchers and machines for trenchless installation, numerous types of attachments to these machines and support equipment have been developed, e.g. [150]:
- attachments to the machines: gravel hopper, conveyor belt, water tank, water sprayer alongside the trench box, blinding device, reel for corrugated plastic pipe and platform for concrete pipes;
- support equipment for auxiliary activities: gravel trailers, attachments for backfill and transport equipment;
- for grade control: laser;

Table 5.2   Comparison of the capacities of a trencher (160 kW) and a V-plough (200 kW) as used for the installation of field drains in various soils and at drain depths ranging from 1.00 m to 1.90 m in the Netherlands [272].

Soil type	Drain depth [m]	Capacity [m hr^{-1}]		Ratio trencher/ trenchless
		Trencher (160 kW)	Trenchless (200 kW)	
Sand:				
	1.00	700	840	1.2
	1.30	600	600	1.0
	1.60	520	430	0.8
	1.90	475	-	-
Clay loam and Clay:				
	1.00	620	1150	1.9
	1.30	540	1050	1.9
	1.60	470	800	1.7
	1.90	420	-	

- for quality control: tracking, rodding, continuous depth recording and video equipment;
- for maintenance: flushing equipment.

Trenchers and machines for trenchless installation combine a number of activities that in the era of manual installation had to be done one after each other, i.e. (i) excavation of the trench, (ii) grade control, (iii) placing the pipes, (iv) placing the envelopes, and (v) blinding (blinding is the first step in backfilling a trench by carefully replacing the excavated soil around and over the drain pipe, mainly to fix the drain pipe in its position). This combination of activities, together with the logistics to guarantee the high speed of installation, imposes additional requirements on materials, operation practices, and quality control, i.e. lightweight and flexible drain pipes, with pre-wrapped envelopes, are required for trenchless machines and recommended for trenchers, although trenchers can also be used with concrete pipes and laser technology for semi-automatic depth and grade control.

These requirements are discussed in the following sections.

5.1.3   Drain pipe materials

Traditionally, clay pipes with a length of about 0.30 m were used. Their production was first mechanized in England and from there it spread over Europe and to the U.S.A. in the mid-19th century [41]. A breakthrough in pipe drainage technology occurred in the 1940s when rigid plastic pipes were introduced in the USA, followed by corrugated PVC and PE pipes in the 1960s in the Netherlands [220]. Nowadays, corrugated PE or PVC is considered to be the preferred standard. The choice depends mainly on the availability of the raw material and its price.

The main advantages of corrugated pipes are the greater length of the pipe, up to 150 m depending on the diameter, and its lower weight per meter. The greater length has reduced the number of joints significantly, reducing the risk of sedimentation and eventually clogging. A large-scale excavation programme, carried out in the Nile Delta in Egypt, revealed that sedimentation was significantly reduced after the introduction of plastic pipes for field drains (Figure 5.1). An excavation programme carried out in India showed similar positive results, with sedimentation in plastic field drains always less than 5 mm [103]. The lower weight of plastic pipes makes transport and handling much easier. Subsequently, small diameter pipes (Ø < 150 mm) were rapidly introduced world-wide.

Concrete collector drains have the same sedimentation problem as the clay pipes. In Egypt sedimentation levels in concrete collector drains, with diameters up to 500 mm, reduced the effective cross sectional area by about 35% [195]. Thus it is not surprising that plastic collector drains perform better than concrete drains [69], mainly because the lower sedimentation rates that offset the higher roughness coefficient caused by the corrugations. However, the introduction of larger diameter plastic pipes (150 < Ø < 300 mm) for collector drains took much longer than the introduction of smaller diameter pipes for field drains, mainly because of the complex manufacturing process [141; 182]. The biggest obstacles that had to be overcome for the introduction of corrugated plastic pipes were: (i) the complex manufacturing process, (ii) making the pipes strong enough and flexible, and at the same time keep the weight per metre low, and (iii) the logistic problems, because plastic pipes are more sensitive to temperature and ultra-violet radiation. Especially when exposed to sunlight, the pipes trends to become brittle [3].

Existing standards were updated to include specifications for the new materials from which the pipes are manufactured. These standards, often originating from countries with a long drainage history, were used to develop international standards, e.g. European Standards [245]. These international standards can be used as a reference for countries without a well-established drainage industry. However, they need to be adapted to specific, local conditions and circumstances, e.g. in Pakistan [251], in India [182] and in Egypt [149].

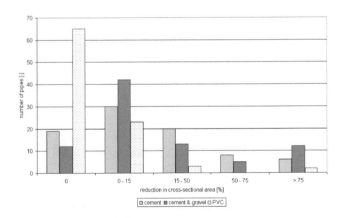

Figure 5.1   Sedimentation in plastic and concrete pipe drains, expressed as a reduction in the cross-sectional area, observed in large-scale excavation programmes in the Nile Delta, Egypt [15]

5.1.4   Envelope materials

A drain envelope has three functions [50]:
- *filter function*: to prevent or restrict soil particles from entering the pipe where they may settle and eventually clog the pipe;
- *hydraulic function*: to constitute a medium of good permeability around the pipe and thus reduce entrance resistance;
- *bedding function*: to provide all-round support to the pipe in order to prevent damage due to the soil load. Note that large-diameter plastic pipes are embedded in gravel especially for this purpose.

These functions are somewhat conflicting as the filter function requires fine envelope materials with small pore sizes and the hydraulic function coarse envelope materials with wider pore sizes. Apart from these conflicting filtering and hydraulic functions, the formulation of functional criteria for envelopes is complicated by a dependence on soil characteristics, mainly soil texture, and installation conditions [247; 281]. Stuyt and Dierickx reviewed the simultaneous development of theory and practical experience in Europe and North America in more detail [246].

Traditionally, the required envelope around the drain pipe consisted of locally available materials like stones, gravel or straw. In arid areas, the technique of using gravel envelopes has been further developed to such a degree that effective gravel envelopes can be designed for most soils [259]. In practice, gravel envelopes are often expensive due to the high transport costs, while their installation is cumbersome and error prone, and requires almost perfect logistic management during installation (Figure 5.2). Moreover, gravel cannot be used when installation is done with trenchless equipment. Subsequently, pre-wrapped envelopes of synthetic material have been under development for some decades. Pre-wrapped envelopes made of artificial fibres are presently used almost everywhere in Europe, in some areas of the United States, and in Egypt.

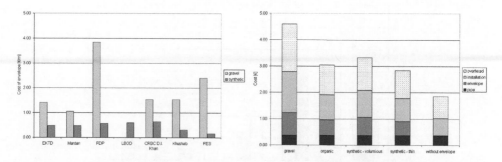

Figure 5.2   Cost comparison between various types of envelopes: (a) total material and transport costs for gravel and synthetic envelopes for various projects in Pakistan; (b) envelope cost as part of the total installation costs in the Netherlands (both 1992 prices) [281].

Since the specifications of envelopes are very soil specific and soils are rather variable, the specifications and effectiveness of envelopes have to be proven in field trials in the areas where they are to be applied [281]. Specialized machines have been developed to pre-wrap sheet and loose-fibre envelopes around the drain pipes, not in the field but in the factory, ensuring a better quality and easier quality control (Figure 5.3).

5.1.5   Quality control

Traditionally it was the farmer himself who installed subsurface drains in his fields. He did it in the off-season using his own labour or engaging a local contractor. Quality control was rather simple as the drains were dug by hand and lines of control were short. Nowadays, in large-scale drainage projects, many persons are involved in the installation process making quality control much more complex. This requires a well designed systematic quality control process: instead of putting the emphasis on post-construction quality checks, it is preferable that the quality control process becomes an integral part of the implementation process, the so-called 'Total Quality System' [150]. In this system, each person in the implementation process, from the planning up to the O&M, is responsible for the quality of his/her own work and for carrying out a quality control on the output of the individual tasks. Basically, if one step is not carried out properly, the persons responsible for the next step should refuse to continue with the process until the previous step has been rectified (Figure 5.4). Two examples of these changes in the quality control process are discussed in more detail.

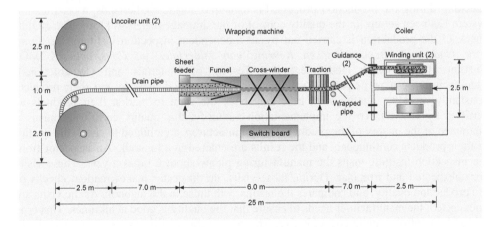

Figure 5.3   Example of an envelope wrapping-unit installed in the EPADP drain pipe factory in Tanta, Egypt. The unit consists of 2 un-coiler units, sheet feeder and funnel, cross-winder and 2 winding units [201].

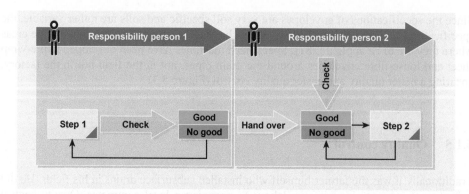

Figure 5.4   Principles of the 'Total Quality Control System' for large-scale subsurface
              drainage programmes: each person at every step in the implementation
              process is responsible for the quality of his/her work [150].

Certification

Traditionally, it was the principal engineer who was responsible for the quality control: he
or his deputy checked the quality of the system during or after the contractor or the
labourers had installed the drain pipes. Certification implies that the responsibility for the
quality is handed over to the manufacturer or the contractor who must guarantee that his
products meet the required certification standards [273]. Certification standards are
deposited with national or international bureaus of standards. The certification is issued
and checked by an independent organization. Control is normally done by random
checking during the production process. For example, in the Netherlands, a certification
system has been set up for the quality control of the drainage materials [150]. The quality
check of the production is carried out by an independent inspection institute: '*Stichting
voor Onderzoek, Beoordeling en Keuring van Materialen en Constructies*/KOMO'
(Institute for Research, Judgement and Testing of Materials and Constructions).
Manufacturers can participate on a voluntarily basis and, if they do and their products
constantly meet the quality standards, they have the right to market their products as
certified by KOMO. To the implementation authority, this quality certificate means a
guarantee of the quality of the product. The manufacturers are obliged to check the quality
of their products continuously and the results are entered in a logbook. An inspector from
the inspection institute visits the manufacturing plants about 6 times a year. These visits
are unexpected and irregular. During these visits, the inspector makes random checks of
the production quality, and compares the results with those in the logbook. As the visits are
unexpected, the manufacturer needs to ensure that the quality is good at all times. This type
of quality control is quite cost-effective. The cost of this certification system amounts to
about 0.5% of total drainage costs, less than half of the costs of post-construction methods
like rodding or continuous depth recording (Table 5.3). Nowadays, the standards for
certification are based on the international standard ISO-9000-series. It should be realized,
however, that certification only covers the quality of product; defects caused by transport,
storage and handling are not included.

Table 5.3   Cost of quality control in the Netherlands expressed as a percentage of the total cost of drainage in large projects [150]

Item	Cost of checking all drains (%)	Intensity of random checks (%)	Cost of random check (%)
Certification of materials[a]			0.5
Rodding	6	15	0.9
Continuous depth recording	50	3	1.5
All three methods			3

[a] Excluding cost of internal quality control by manufacturer.

Laser control

Installing a drain or collector pipe at the proper grade (slope) is essential for the functionality of the drain [44; 150]. Traditionally, this was done manually during or immediately after installation by measuring the level of the top of the drain pipe every 5 m.

The high speed of installation and the introduction of the trenchless drainage machine let to the development of laser control. Laser equipment for drainage basically consists of two components: (i) a transmitter, which is positioned in the field, and (ii) a receiver mounted on the trench box of the drainage machine (Figure 5.5). The receiver is electrically connected to the hydraulic system of the lifting cylinders of the trencher and is programmed in such a way that it automatically adjusts the depth of the trench box to the preset grades stored in the memory of in the transmitter. There are also indictor lights on the operator's display (receiver display) so that he can check the system continuously. Laser control has greatly improved the quality of installation (Figure 5.6).

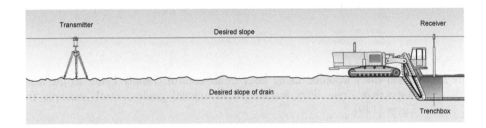

Figure 5.5   Schematic set-up of the laser equipment: the transmitter in the field and the receiver on the trench box of the drainage machine.

5.1.6   Organization

Large-scale drainage projects are complex and numerous stakeholders are involved. The stakeholders are the farmers, national or provincial government organizations, planning and implementation authorities, drainage contractors, suppliers of drainage materials and machinery.

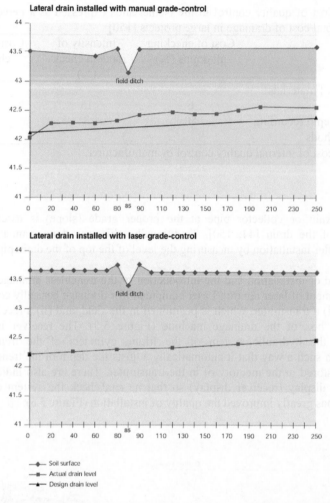

Figure 5.6   Drain lines installed using laser grade-control comply better to the design slope than drain lines installed with manual grade-control: example from Egypt [149].

All these stakeholders have their own specific interest. The implementation process can be divided in four main steps [150] (Figure 5.7):
- policy preparation and decision-making;
- technical, organizational and administrative preparation;
- actual implementation: field investigations, design, planning and budgeting, tendering and construction;
- handing-over and O&M.

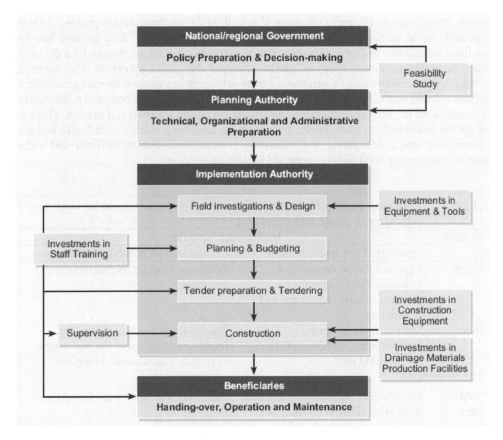

Figure 5.7   Main activities and players in the implementation process [150]

For each step in the implementation process, one authority will have the overall responsibility, e.g. the national or regional government in the policy and decision-making process; a planning authority in the preparation; an implementation authority for the actual implementation; and, of course, the farmers or their representatives to operate and maintain the system. Most stakeholders, however, are also involved in the other steps in the implementation process. Their roles and responsibilities are described in the implementation mode that defines which activities are carried out by whom and under what conditions. Basically two implementation modes are possible to carry out activities, i.e.:

- by a specialized government entity, for example EPADP in Egypt [149];
- contracted to a specialized company, for example the consortia formed in Pakistan under the authority of WAPDA [150].

The implementation authority, which can be a public (national or regional government) or private (a group of farmers) organization, has to decide the mode of implementation for each phase of the installation of a subsurface drainage system. In many cases the mode is already routinely prescribed by the rules and regulations of the country and/or financers. In countries with a well-developed drainage tradition and drainage industry, contractors and/or consultants usually carry out most, if not all, of the implementation process. This is especially so in countries where privatisation is well established. If a country has no qualified consultants and/or contractors and/or suppliers of drainage materials, a decision must be made to either fully or partly privatize the development of the drainage technology, i.e. contractors / consultants / material supply, or request special government entities to build up the knowledge and skills and/or purchase the equipment. An alternative is to obtain all the services, equipment and materials on the international market. There is no golden rule which mode to apply, both methods have been used: e.g. in Egypt and the Netherlands specialized public authorities were established, but in Pakistan and India special project organizations were created (Table 5.4).

Table 5.4   Examples of implementation modes used in some countries with major subsurface drainage activities [150]

Country	Implementation mode	
	Specialized government entity	Contracted to a specialized company
Egypt	Egyptian Public Authority for Drainage Projects for the planning, design, tendering and contracting	Private and public contractors for the installation
India	State Command Area Development Authority (CADA) for planning etc.	Special project organizations for implementation under the authority of CADA in Rajasthan and Haryana
The Netherlands:		
'Old' lands	Government Service for Land and Water Use (GSLWU) for planning, preparation and supervision	Public contractors for the installation
'New' polders	IJsselmeerpolders Development Authority (RIJP) for the reclamation and development of new lands[a]	Public contractors for the installation
Pakistan	Central & Provincial Governments for the for the planning, design, tendering and contracting	Special project organizations under the authority of the Provincial Governments in Sindh Province, Northwest Frontier Province and Punjab
USA [163]	Corporate or mutual drainage undertakings by two or more landowners cooperating without special organization under State drainage laws Legally organized public drainage organizations administered by public officials.	

[a] The tasks of the RIJP came to an end in 1996 with the completion of the Southern Flevoland polder. The organization merged into the Regional Directorate IJsselmeerpolders of the Directorate General of Public Works and Water Management.

5.1.7   Capacity Building

To implement all these innovations in drainage equipment, materials, installation techniques and procedures requires that all persons involved are properly trained. Especially when the implementation is contracted to a specialized company, the implementation authority needs to specify that only certificated staff is employed. Next to formal education and training to obtain basic knowledge, experience with large- scale projects shows that for the practical skills and procedures on-the-spot training is the most practical and effective approach [205]. The approach is based on the principle that the trainers go to the field instead of the field staff going to the trainers. An example of this approach is the Drainage Training Centre (DTC) in Tanta, Egypt, established in 1991 [149; 150]. DTC is the result of a long-term co-operation between RIJP and EPADP. Those two organizations cooperated in the Drainage Executive Management Project (DEMP). When the DEMP project commenced there was no former training programme. The project started with training of EPADP staff in the Netherlands. At the same time, Dutch instructors together with their Egyptian counterparts started to visit and train the staff of EPADP and the contractors in the directorates all over Egypt. This training programme was known as in-service training, and became a regular event. Gradually, the Egyptian instructors took over the training ('train the trainers').

The in-service training proved to be an instrument not only to train staff successfully in mechanized drainage implementation, but also to introduce new techniques for quality control, such as using laser equipment and rodding equipment. After some years it was felt, however, that the range of training was still too limited. The need for more specific training courses became evident and the visits of EPADP staff to vocational training centres in the Netherlands convinced the EPADP management of the need for a permanent training centre in Egypt. This led to the establishment of DTC. The training activities at the DTC focus on personnel of the EPADP organization and contractors, in order to:
 • increase their skills for the job;
 • obtain essential knowledge to perform their job;
 • improve the quality and the quantity of their performance.

DTC has all the facilities to conduct practical training courses. Besides the theoretical lessons much attention is paid to practical training of the trainees. All the instructors at the DTC are engineers with many years of experience in drainage practice in Egypt. The annual training programme includes: field engineer execution courses, maintenance engineer courses, laser courses, surveying courses, operating drainage machines courses, channel maintenance with mowing buckets and so forth. This in-service training has become an integrated part of DTC's course programme. That Egypt has nowadays one of the largest and most modern subsurface drainage programmes in the world can, to a large extend, be attributed to these capacity building activities. This is especially remarkable, because the developments in Egypt took place in a relatively short period: over a period of 40 years manual installation practices were almost completely mechanized, including the introduction of new materials (Figure 5.8).

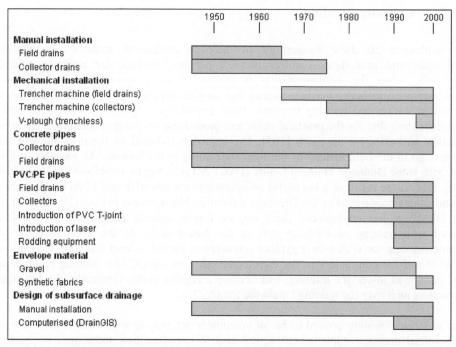

Figure 5.8   Technical developments in the large-scale implementation of drainage projects in Egypt [149].

5.1.8   Conclusions

Subsurface drainage is a form of drainage that was widely introduced in many parts of the world in the second half of the 20th century. This was only possible because the prevailing empirical knowledge of drainage and salinity control gained a solid theoretical footing in the first half of the 20th century. This introduction was further accelerated by the rapid developments in mechanized installation from the 1940s onwards. To make the shift from manual to mechanized installation a number of problems had to be solved. New drainage materials were developed to replace the traditional drain pipes made from clay or concrete and drain envelopes made of organic materials or gravel. Plastic drain pipes, made of PVC or PE and synthetic envelopes resulted in lower transportation and installation costs because of the weight and shape of the materials and better quality of construction. A shift from post-construction quality control to a total quality control system enables achieving a high quality of the installed systems even with the ever-increasing speed of installation. This was only possible because at the same time, new modes for the implementation process were developed and implemented. Nowadays, basically two modes are used: implementation by a (specialized) government entity or contracted to a specialized drainage company. The choice mainly depends on the existing organization structure in a country. Last but not least, the introduction of modernized drainage machinery and installation techniques can only be successfully achieved if all people involved in the implementation process are properly trained. In addition to formal education and training, the so-called 'in-service training' approach, based on the principle that the trainers go to the field instead of the field staff going to the trainers, proved to be a successful method.

That all these improvements paid off is well illustrated by another quotation of Van Schilfgaarde, who twenty years after his call (in 1979) to develop practical and convenient installation methods, concludes that *'the systematic planning and design of drainage systems has rapidly changed with tremendous improvement in drainage tubing, machinery and methods of installation, drainage envelopes (natural and manufactured), and techniques for quality assurance and control'* [229]. It is my belief that these efforts to improve subsurface drainage practices will facilitate the further introduction of pipe drainage in the world and through this contributes to a better, more sustainable, use of the world's precious land and water resources. However, further research and development is still needed to meet the specific needs of the emerging and least developed countries each with its specific climatic, physical and social conditions. Furthermore, the specific needs of drainage are also changing, particularly with regards to the quality of drainage water. This also requires changes in the drainage system design and corresponding installation practices.

5.2   The added value of research on drainage in irrigated agriculture[11]

5.2.1   Introduction

To make investments in subsurface drainage cost-effective and sustainable research plays an essential role. That research has its value for money is quite well recognized in the Western World. For instance, in the Netherlands research organizations have been privatized[150] [150] and in Australia returns on investments in research are closely monitored and prove to be very positive [216]. Research in emerging and the least developed countries, however, always has to prove itself and continuous support is generally lacking. Despite the fact that the results, i.e. more efficient and effective subsurface drainage systems, are appreciated by international donors [287; 297]. This section shows that applied research on drainage in developing countries has its 'value for money' and that the benefits resulting from research easily outweigh its costs, especially when research is linked to drainage implementation. The analysis is based on over 40 years of partnership in applied research between research institutions in Egypt, Pakistan, and India with Alterra-ILRI of the Netherlands. The tangible research results are presented and whenever possible their impacts are translated in actual or potential savings. The impacts and savings are presented for the four main steps of subsurface drainage practices, i.e. identification, planning and design, installation, and O&M.

[11] Published as: Ritzema, H.P., Wolters, W., Bhutta, M.N., Gupta, S.K., Abdel-Dayem, S., 2007. The Added Value of Research on Drainage in Irrigated Agriculture. *Irrigation and Drainage*, **56**, S205 - S215. DOI: 10.1002/ird.337.

5.2.2   Identification of the need for subsurface drainage

Measuring soil salinity with the EM38

The traditional method to assess the soil salinity status of the soil is through soil sampling followed by laboratory analysis. This is a labour-intensive and expensive method, e.g. in the fiscal year 1993/1994, the SCARP Monitoring Organization in Pakistan collected about 3,000 samples at a cost of € 1.25 per sample (Note: In this section all prices are converted in 2006 prices: € 1.00 ≈ US$ 1.20). To reduce costs and to increase the accuracy a new instrument for measuring soil salinity through electromagnetic induction, the EM38, was tested in Egypt, India and Pakistan. The results show that the EM38 can be used in pre-drainage investigations, in monitoring the performance of drainage systems, and to assess mitigating measures when problems arise during O&M. The costs of using the EM38 are substantially lower than those of the traditional method [281]. Furthermore, the quality of the measured salinity data is better as the instrument measures larger soil volumes and the measurement is direct and fast [144].

Criteria for upgrading

The economic lifetime of subsurface drainage systems varies between 25 and 30 years. Criteria are needed to assess when the O&M costs become so high that it is better to rehabilitate or to replace a system. In Egypt, research shows that a single indicator is not sufficient to assess whether a subsurface drainage system is in need for upgrading. Instead, a combination of indicators should be used, i.e. age of the system, number of complaints, depth of the water table and maintenance cost [63]. A three-step performance assessment methodology was developed to use these indicators. The decision to initiate a upgrading project is taken only when the results of the three steps confirm a need. As each step is only undertaken when the previous step has confirmed its necessity, considerable savings are obtained.

Improving surface drainage to reduce cost of subsurface drainage

Contrary to irrigation canal systems, surface drains are not self-cleaning, restricting the existing capacity of these drains. The case studies from Pakistan (Section 4.3) show that investments in maintenance of the surface drainage systems, which are generally low compared to investments in the implementation of subsurface drainage, can significantly reduce the need and/or capacity of subsurface drainage.

Interceptor drains to minimize drainage requirements

Seepage from irrigation canals has long been considered one of the major contributors to waterlogging and salinity. The case studies from Pakistan show that the effects of interceptor drains do not justify the large investments involved (Section 4.3). The implementation of unnecessary, ineffective and costly interceptor drains in the FESS project could be prevented to the tune of over € 8.5 million [110]. As well, an estimated annual cost of € 0.8 million for ineffective re-circulation of water, due to the low percentage of net seepage interception, was avoided [291].

Lining of irrigation canals to reduce seepage

An option to reduce seepage is lining. Research conduced in Pakistan shows that the impact these seepage rates have on subsurface drainage systems is, however, negligible (Section 4.3). The research resulted in the cancellation of the lining for the Malik Branch, saving about € 8.3 million [291]. The research on lining in Pakistan led to improvements in a lining programme in China (Tarim II Basin, Xinjiang Province) where a small extra investment in lining led to great saving of water for environmental purposes and reduction of waterlogging [292].

Groundwater modelling to identify areas in need for subsurface drainage

In large-scale irrigation projects, subsurface drainage is in general installed in the total project area. Groundwater modelling in Pakistan showed that subsurface drainage is generally not needed everywhere and that design rate could be reduced (Section 4.3). Had this research result been known before the installation of FDP, about € 2.9 million could have been saved. The result also had a major impact on further development of the FESS project.

5.2.4   Planning and design

Modified lay-out for areas with rice in the cropping pattern

In Egypt, rice is cultivated in rotation with 'dry-foot' crops. The implementation of conventional free-flowing subsurface drainage systems serving a mixed pattern of crops including rice caused excessive drainage from the rice fields. To reduce water losses from areas cultivated with rice without restricting drainage from other crop areas, a modified layout of the subsurface drainage system was developed [66]. A monitoring programme showed that farmers adjusted themselves nicely to the system and managed to use 43% less irrigation water, saving the same percentage on pumping costs [63]. In the Nile Delta, where annually about 0.4 Mha is cultivated with rice, the potential saving would be about € 10 million. Furthermore, the design discharge rate for collector drains (4 mm d^{-1}) could be reduced to 3 mm d^{-1}, the design discharge for non-rice areas. This resulted in smaller pipe diameters, thus savings in investment costs.

Controlled drainage

Most subsurface drainage systems in irrigated lands have free-flow outfall conditions. One of the original design assumptions, i.e. a high water table, however, only occurs for short periods and therefore most of the time excessive drainage occurs. To investigate whether controlled drainage can maintain higher water tables without increasing soil salinity the simulation model DRAINMOD-S was applied for the western Nile Delta. The results showed that controlled drainage has the potential to maintain and even increase yields while increasing irrigation water use efficiency by 15 to 20% [284]. The savings can be obtained with low-cost and easily operated devices. Similar results were found in controlled drainage experiments in India [103]. Controlled drainage also reduces the downstream environmental impacts as the total salt load is reduced proportionally with the water savings.

Design discharge rate

Various studies conducted to verify drainage design criteria in pilot areas in Egypt, India and Pakistan have shown that in general the design criteria are too conservative and can be reduced. In Egypt, a design discharge rate of 0.9 mm d^{-1} is sufficient to cope with the prevailing losses of irrigation water and to maintain favourable soil salinity levels [14]. This is 10% lower than assumed in the design. Subsequently, the design discharge rates for collector drains can also be decreased [68]. In India, research shows that the original design rate for salinity control (2.0 mm d^{-1}) can be reduced to 1.0-1.5 mm d^{-1} [182; 203]. In Pakistan, field monitoring programmes and computer simulations, indicate that the field drainage design discharge can be reduced from the initial value of 3.5 mm d^{-1} to 1.5 mm d^{-1} [291].

Design drain depth

The same applies for the depth of the drains. In Egypt, pilot area research showed that a design depth of the water table of 0.80 m proved to be sufficient [14]. The most cost-effective way to obtain this depth at the given discharge is to install drains at a depth between 1.20 to 1.40 m [149]. In India, field data combined with simulations using SALTMOD indicate that drain depth, under gravity flow conditions, can be reduced to 0.9 to 1.0 m [244]. In Pakistan, the design depth is also gradually reduced: from a drain depth of 2.25 - 2.40 m in the 1980s to 1.50 - 2.10 m in the 1990s [177]. Reducing drain depth does not so much result in major savings in the cost of implementation, the savings are more environmental. Salts from the deeper subsoil are not disturbed, resulting in lower salt load in the drainage effluent.

Automation of the planning and design process

In the 1960s and 1970s, at the start of the large-scale implementation programmes, designs were made by hand. Since then, computerization of the design process was gradually introduced; e.g. in Egypt [149], in India [182] and in Pakistan [251]. The automation has improved the quality of the whole planning and design process. It increased the pace of implementation and reduced costs, especially of large-scale projects.

Use of simulation models

The introduction of simulation models in the 1980s has greatly improved the knowledge of the functioning of subsurface drainage, design of these systems, and analyses of water and salt movement under varying and complex field conditions as in the case of integrated irrigation and drainage management [5].

5.2.5   Installation

Pilot area research to test installation methods and materials

Numerous research activities were conducted to develop and test new drainage materials and installation methods [150; 201]. The savings are difficult to quantify in monetary terms but impacts on pace and quality of construction were huge. For example, the introduction of plastic pipes in Egypt has increased the installation rate by an estimated

20%, thereby reducing installation cost with the same percentage [265]. The introduction of plastic pipes also significantly reduced sedimentation [15]. This could only be achieved by improved grade control through the introduction of laser (Section 5.1.5). The use of synthetic envelopes in the FDP in Pakistan alone would have given savings around € 1.5 million per year [291]. The combined mechanical/manual installation method in India has lowered the installation costs and at the same time increases employment of the poor [103]. The introduction of trenchless drainage in Egypt reduces the installation cost per hectare by about 18% with estimated savings of about € 2.25 million per year [98]. That these research activities had their added value is best illustrated by the fact that Egypt and Pakistan have nowadays some of the largest, most modern and effective subsurface drainage programmes in the world with considerable cost savings.

Improvement of installation practices

Next to the new materials and methods, installation practices were further improved with the introduction of improved methods for quality control, like rodding and video inspection [150]. Training and capacity building programmes were set up to introduce all these innovations in drainage equipment, materials, installation techniques and procedures. For example, in Egypt DTC was established in Tanta [150]. Outputs of operational research programmes are used to improve planning and implementation practices (Figure 5.9).

Organization

Specialized organizations were created to implement the large-scale drainage projects [150; 201]. In Egypt, installation is done by public and private contractors employed by EPADP. In India, and Pakistan, consortia under the authority of Provincial or State Governments were formed. The success of the implementation mode adopted in Egypt can be demonstrated by the high implementation rates of subsurface drainage system projects. Annually about 63,000 ha are provided by new subsurface drainage systems while old drainage systems are upgraded in about 12,.600 ha. The pace of implementation is slower in Pakistan and especially India. In India, however, the tide is changing. A private entrepreneur is undertaking a drainage project in about 3,500 ha in Maharashtra and Karnataka with funding from the Central, State Government and stakeholders in the ratio of 60:20:20. Banks provide loans to the stakeholders.

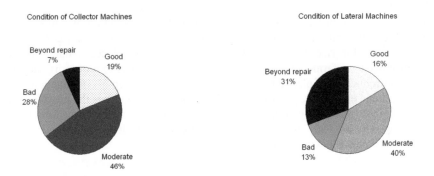

Figure 5.9   Operational research on the performance of drainage machines greatly improved the planning: example of operational research in Egypt [142].

Farmer's participation

Farmer's participation is gradually increasing. In this difficult and time-consuming process research plays an important role. Examples are farmer's participation in the design, implementation and O&M of a subsurface drainage system in Pakistan [21]; the participation of farmers in the operation of the modified system in Egypt [63] and their growing role in the newly established Water Boards; and, the participation of farmers, including women, in subsurface drainage activities in India. Financially, farmers in Egypt pay back the full investment cost of subsurface drainage over 20 years, without interest. Next to the monetary benefits, the socio-economic benefits, e.g. in terms of labour opportunity, increased income, improved position of women, landless and tenants, are also high [122]. Thus value for money is not only monetary as was illustrated with the example from Uppugunduru Drainage pilot area in Andhra Pradesh, where subsurface drainage decreased the workload of especially the women in rice cultivation practices (Figure 3.7) [103].

Advisory Panel

In 1975, the Egyptian Government established the joint Egyptian-Dutch Advisory Panel on Land Drainage (APP) with the aim to improve the implementation of its subsurface drainage programme [19]. The panel members are high-level Egyptian and Dutch administrators, scientists and consultants. The Panel initially focussed on technology and design criteria of land drainage then gradually moved its attention to water management. The Panel acts as a think-tank for policy making and strategic planning. The Panel stands as a unique model of bilateral cooperation. This is best illustrated by a quote of Dr. Abu Zeid, the Egyptian Minister of Water Resources and Irrigation, who in an interview with a Dutch newspaper (NRC Handelsblad 28-01-2007) said *'The Netherlands are not our only partner, but without doubt the most important and most effective one. Without, we would never have reached our current level'*.

5.2.6   Operation and maintenance

Improved O&M practices

Research helped to improve O&M practices through the introduction of improved design concepts. E.g., the introduction of the modified drainage system in Egypt (see Case Modified lay-out for rice areas) not only reduced operational costs, but also reduced maintenance needs as farmers no longer illegally blocked drains to reduce irrigation water losses. The introduction of plastic field and collector drains, often in combination with pre-wrapped synthetic envelopes, greatly reduced sedimentation and thus the need for flushing [103; 195]. Improved flushing equipment and methods to remove sediment from the drains were developed [149; 150]. Increased farmer's participation led to more ownership and less misuse or illegal blocking [63].

5.2.7   Conclusions

Drainage, as a tool to combat waterlogging and salinity, plays a major role to safeguard investments in irrigation, to promote economic growth, and to ensure the sustainability of irrigated agriculture. To make sure that investments in drainage are sound and sustainable, countries like Egypt, India and Pakistan invest heavily in research on drainage-related water management. That these research efforts have their value for money is illustrated by the fact that in these countries some of the largest, most modern and effective subsurface drainage programmes are implemented. This was achieved in combination with considerable cost savings. An achievement that only could be realized by linking research with design and implementation practices.

Over the last 40 years, applied research activities have helped to modernize subsurface drainage practices and considerable savings have been achieved by the introduction of: (i) new methods to investigate and identify areas in need of drainage, (ii) new design and planning methods, (iii) new materials for pipe drains and envelopes, (iv) improved drainage machinery and equipment, and (v) improved installation, O&M methods and practices. Last but not least, research has helped to improve the organization of subsurface drainage operations and institutions. All these improvements could be achieved because these countries not only invested in research but also in training all personnel involved in applying the new and innovative practices. The examples discussed in this section (summarized in Table 5.5) show that countries do well in attaching research to large-scale implementation.

Research in irrigated agriculture is, however, not yet over and continuous support will be needed. The introduction of new crop varieties and crop diversification will require improved water management practices. Aging systems require new methods for upgrading. The hydrological environment is also changing: in many regions, less water is available due to increasing scarcity and upstream use. In other regions, water quality deteriorates or the discharge of drainage effluent is restricted. Furthermore, one of the effects of climate change is that extremes are occurring more often: both wet and dry periods are on the increase. Finally, socio-economic conditions are also changing, asking for a participatory approach. To sustain irrigated agriculture under these changing conditions requires a continuous support in drainage research; in this section, it has been shown that this research has its value for money.

Table 5.5   Summary of the benefits and savings due to research

Research finding	Country[a]			(Potential) impact
	Eg	In	Pa	
Identification:				
• Measuring soil salinity with EM38	x	x	x	Financial benefits hard to quantify, but substantial improved quality of monitoring
• Criteria for upgrading	x			Better planning of upgrading works
• Improving surface drainage			x	Lower investments costs
• Interceptor drains to minimize drainage requirements			x	Savings of € 8.3 million in the FESS project only.
• Groundwater modelling to identify the areas in need for subsurface drainage			x	Saving of € 2.9 million in the FDP project and considerable impacts on the FESS project
• Lining of irrigation canals to reduce seepage			x	Savings of € 8.3 million in the FESS project only
Planning and design:				
• Modified lay-out for area with rice in the cropping pattern	x			Savings of € 10 million in irrigation water and pumping costs and a 25% reduction in the design discharge rate
• Controlled drainage	x	x		Irrigation water efficiency is increased by 15 to 20%
• Design depth	x	x	x	Lower design drain depths result in lower installation costs
• Design discharge	x	x	x	Smaller pipe diameters and thus lower material costs
• Automation of the design process	x	x	x	Better systems
Installation:				
• Plastic drain pipes	x			20% increase in installation rate
• Synthetic envelopes			x	Savings of € 1.5 million per year
• Trenchless drainage				Potential saving of € 2.25 million per year
• Improvement of installation practices				Savings tens of millions EURO.
• Organization	x	x	x	As above
• Advisory panel	x			As above
Operation and maintenance:				
• Modified system for rice areas	x			Savings of € 10 million in pumping costs and maintenance
• Flushing	x			33% cost reduction
• Farmer's participation	x	x	x	10% savings in labour cost

[a] Eg = Egypt; In = India; Pa = Pakistan

6   Capacity development to improve subsurface drainage practices

6.1   An integrated approach for capacity development in drainage[12]

6.1.1   Introduction

Capacity development is an essential element to achieve improved irrigation and drainage practices. Capacity development aims to develop institutions, their managerial systems, and their human resources to make the sector more effective in delivery of services [256]. Within the framework of IWRM, capacity development has to focus on three elements [83]:

- creating an enabling environment with appropriate policy and legal frameworks;
- institutional development, including community participation;
- human resources development and strengthening of management systems.

Thus capacity development addresses three levels: the individual, the institution and the enabling environment. Each level has different goals, activities and outputs [266]. In this respect, capacity development is as much a process as a product [119]. In this process, the more concrete or explicit aspects of capacity development such as training and institutional strengthening have to be linked with the local or tacit knowledge and aspects of ownership. Luijendijk and Mejia-Velez define explicit knowledge as the knowledge that *'can be expressed in facts and numbers and can be easily communicated and shared in the form of hard data, scientific formulae, codified procedures, or universal principles'* and tacit knowledge as *'highly personalized and hard to formalize, subjective insights, intuitions and hunches'* [133].

Although the principles of IWRM are generally accepted, it is less clear how they should be applied. Since 1995, funding for capacity development in the water sector has grown ten-fold, but despite this increase the pressure on water resources has not diminished [256]. Two principal weaknesses in capacity development are contributing to this. Firstly, capacity development activities are often geared towards the past rather than the future and they are not coupled to changes [248]. Consequently, new skills and attitudes are unlikely to be used, as organizations are remarkably stable and new insights acquired from training alone are unlikely to change things. Secondly, capacity development activities tend to focus on explicit knowledge and to neglect tacit knowledge. Boon [132] estimated that this tacit or undocumented (local) knowledge accounts for 75 to 95% of the total organizational knowledge. Thus to increase the impact of capacity development activities, the challenge is to link tacit with explicit knowledge and to update it continuously. In this section, an approach to achieve this challenge is presented based on three elements: research, education, and advisory services.

---

[12] Published as: Ritzema, HP, Wolters, W, and Terwisscha van Scheltinga, CTHM. 2008. Lessons learned with an integrated approach for capacity development in agricultural land drainage. *Irrigation and Drainage*, **57**, 354–365. DOI: 10.1002/ird.431.

6.1.2   Materials and methods

The capacity development approach presented in this section aims to improve and sustain land and water resources in emerging and least developed countries with the focus on practical implementations of drainage, irrigation and related water management in farmers' fields. Not only the development of new knowledge is the prime focus, but also the translation and adoption of knowledge developed elsewhere. In this approach three routes are followed [205]:

- training programmes for mid-career land and water management professionals;
- joint projects in applied research and institutional development through on-the-job cooperation on technical, social and institutional issues with the focus on finding practical and directly applicable solutions. These projects are conducted in association with research organizations in the region or country where the problems are occurring;
- dissemination of knowledge through publications and papers. By working together in joint-projects and in training activities, the trainers also benefit by picking up new skills and knowledge. They use this knowledge to update the explicit knowledge which subsequently is disseminated through lecture notes, handbooks, special reports, papers, etc.

In this section, five examples are presented how, through linking explicit with tacit knowledge, the capacity development process is enhanced, i.e.:

- capacity building through training and dissemination of knowledge;
- capacity building to improve subsurface drainage practices in Egypt;
- capacity development to combat waterlogging and salinity in irrigated lands in India;
- capacity development to improve drainage practices in Pakistan;
- capacity building for the sustainable management of tropical peatlands in South-east Asia.

The five examples address the five strategic phases identified by the ICID Working Group on Capacity Building, Training and Education (WG-CBTE) (Figure 6.1). In the examples, capacity development was generally only one of the objectives of the projects and subsequently not all five steps were always addressed equally. In the discussion it will be shown that, because capacity development is as much a process as a product, it is not always required to include all steps.

In the discussion, the approach as presented in this section will be compared with the approach developed by Nonaka and Takeuchi [153]. In this approach, which Nonaka calls 'knowledge creating process', the capacity development process is divided in four phases (Figure 6.2):

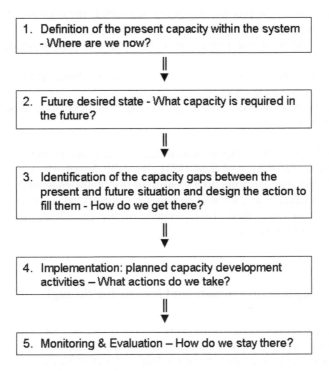

Figure 6.1   The five strategic steps of capacity development in irrigation and drainage [109]

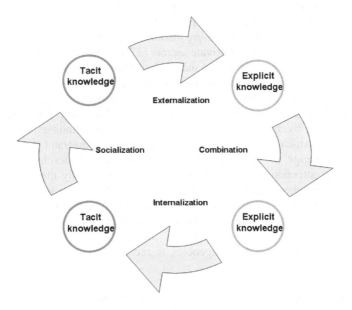

Figure 6.2   The knowledge creating process (after [153])

- *socialization* or the process of creating new tacit knowledge out of existing (tacit) knowledge by sharing experiences;
- *externalization* or the process of converting this tacit knowledge in explicit knowledge;
- *combination* or the process to convert explicit knowledge into more complex and systematic sets of knowledge;
- *internalization* or the process of turning this explicit knowledge into tacit knowledge.

Thus tacit knowledge supports explicit knowledge and becomes a synonym for 'capacity to act' or competence to solve problems [133]. We will describe how we link tacit and explicit knowledge and how the four phases of the knowledge creation process are addressed. In the conclusions, we will summarize the essential elements of the capacity development approach presented in this section, i.e. the integration of applied research, education and advisory services.

6.1.3   Training and dissemination of knowledge through publications

Since 1962, Alterra-ILRI organizes mid-career training courses for professionals in land and water management [108]. The focus is practical, much emphasis is given to exercises and study visits as can be seen from the study load: lectures 45%, exercises 35% and study visits 20%. In exercises, participants have to use their own tacit knowledge and during study visits, they can experience how others use theirs. To enhance the socialization phase of the knowledge creating process, overlap in lectures has been created: the same subjects are treated by lecturers with different background to ensure that the participants are exposed to the possibilities and limitations of the engineering options from various angles. To guarantee a good mix of theory and practice, lecturers are recruited from different organizations: from universities and training institutes (25%), from research organizations (37%) and from the government and private sectors (37%). The introduction of a problem-oriented approach enables participants to match their tacit (or local) knowledge with the explicit and tacit knowledge presented by the lecturers. In group assignments, participants work together on case studies. They are encouraged to exchange their tacit knowledge and to match it with the explicit knowledge presented in the assignment. The case studies are continuously updated by the lecturers using experiences and findings of research and advisory projects in which they are engaged. To address the constraint that organizations are reluctant to change, a separate module on institutional development has been developed. Special attention is paid to how participants can apply their newly acquired skills and knowledge when they return to their normal work, i.e. the internalization phase in the knowledge creating process. To assess the effects of the courses, each course is evaluated by the participants. The evaluations are used by the courses advisory board, which consists of representatives from Dutch universities, government agencies, the private sector and international donor agencies, to recommend changes in the curriculum and lecturers.

To give alumni the opportunity to reflect on current developments and to increase and exchange their knowledge and experience, refresher courses are organized since the mid

1990s. Some of these courses have even evolved in regular training courses, e.g. at the Acharya N.G. Ranga Agricultural University (ANGRAU), Hyderabad, Andhra Pradesh, India, a special Water Technology Centre has been established for this purpose [25].

The tacit knowledge emerging from the courses is transformed in explicit knowledge by updating the lecture notes. These lectures notes are published to make them available to a wider audience: not only to the participants and alumni, but also other professionals working in land and water management. The main characteristics of these publications are that they are 'practical' which means that they can be used in the day-to-day work. Similar to the courses, the publications present a mix of theory and practices. An example is the publication '*Drainage principles and applications*', known as '*Handbook 16*', first published in the 1970s. As the transfer from tacit knowledge into explicit knowledge is a continuous process, the publication has been updated and revised a number of times [191; 192].

6.1.4   Capacity development to improve subsurface drainage practices

In 1975, co-operation with Egypt started with the aim to assist Egypt with the large-scale implementation of subsurface drainage. The co-operation was done along three lines: (i) supporting policy making and strategic planning, (ii) strengthening research, and (iii) improving implementation.

The Egyptian-Dutch Advisory Panel (APP) supports the Egyptian Minister of Water Resources and Irrigation with policy making and strategic planning In the period 1976-2004, the Panel has initiated 42 projects in the field of land drainage, reuse of drainage water, hydrology, groundwater management, maintenance, planning, institutional development and training.

To strengthen research capacity, co-operation with DRI was established. Initially, the emphasis was on technical cooperation gradually changing to institutional support [63; 64]. These changes were guided and advised by the Panel. To improve the field, laboratory and desk-top research methodologies, in-service training was provided by the long-term experts, supplemented by training and study tours abroad. For example, in the Drainage Technology and Pilot Areas (DTPA) project, 43 staff members of DRI followed training courses organized in Egypt (total 108 person-months), 21 staff members followed a training course or study tour abroad (23 person-months), 10 staff members obtained their MSc and 6 their PhD.

To improve the actual implementation of drainage projects, cooperation between EPADP and the Dutch Directorate General of Public Works and Water Management/*Rijkswaterstaat* was established (Section 5.1.7). Next to formal education and training, on-the-spot training was provided and DTC was established. The target group consisted of professionals working for EPADP and the contractors: from drainage machine operators to field engineers.

6.1.5   Capacity development to combat waterlogging and salinity

In India, drainage and drainage-related water management problems are much more diverse than in Egypt. Beside the agro-climate differences, the social and economic settings also vary per state. In 1984, capacity development activities to support the development of subsurface drainage practices to combat waterlogging and salinity in irrigated lands were initiated at two levels. At the national level, the cooperation with CSSRI at Karnal focussed on capacity development in modelling and concept development. At the state level, cooperation with the state agricultural universities in Andhra Pradesh, Gujarat, Karnataka and Rajasthan focussed on modification of these models and concepts to local conditions and assistance with their implementation. A step-wise approach in capacity development was adopted [102]:

- basic training was provided by sending project staff to regular training courses at specialized centres in India and abroad;
- tailor-made courses were organized to train project staff and selected university staff on more advanced subject matters;
- project expert meetings were organized, in rotation at one of the universities, to exchange experiences and to synchronize research activities;
- collaborative research programmes at universities and research organizations abroad were organized to provide in-depth training on specialized subjects.

These four training activities were supplemented by in-service training to introduce new research methods and tools. Study tours were organized to acquaint senior project staff with practices from elsewhere and to disseminate the knowledge and experiences obtained in the project to a broader audience by participation in international workshops and symposia. Finally, to disseminate the knowledge and experiences within India, a training centre was established at CSSRI.

In the period 1996 – 2002, 113 capacity development activities were conducted, in which 388 persons participated, in total 335 person-months (Table 6.1). Using the newly developed knowledge, six pilot areas in farmer's fields, one experimental plot and one large-scale monitoring site were established to develop site-specific subsurface drainage recommendations for the various agro-climatic regions in India [203]. The results of the research activities were published in numerous technical reports, papers published in national and international journals, conference proceedings and MTech and PhD theses.

6.1.6   Capacity development to increase farmers' participation

In Pakistan, cooperation with IWASRI initially followed the same approach as in India but the focus gradually shifted from capacity development aiming to improve the technical aspects of drainage design and implementation to capacity development to increase farmers' participation. To develop these skills new elements were introduced [111]:

- social impact assessments in large-scale drainage systems;
- action research in farmers' implemented drainage systems;

Table 6.1   Capacity development activities conducted in India in the period 1996-2003
              [102]

Type of Capacity development Activity	No. of courses	No. of participants	Person-months trained
Regular training courses in India	6	24	22
Regular training courses abroad	8	28	63
Tailor-made courses in India	12	144	86
Project expert meetings in India	4	57	11
Collaborative research programmes abroad	28	28	69
In-service training of visiting consultants	40	17	20
Study tours abroad	12	29	13
National Training courses	3	61	51
Total	113	388	335

- advice to national services and consultants on methods to involve farmers in the planning, implementation and O&M of drainage systems;
- organization of 'Expert Platform Meetings' on participatory methods in drainage;
- inventory of available expertise on non-technical issues related to farmers' participation in drainage.

With the assistance of an NGO, Action Aid Pakistan (AAPk) and in close cooperation and with the assistance of the farmers, a subsurface drainage system was installed in a site of about 100 ha in Fordwah Eastern Sadiqia [21]. To train the farmers a three-step approach was followed. The first step was to develop tailor-made training modules on participatory drainage. To get a good mix between tacit and explicit knowledge, a series of national expert consultations on farmers' participation in drainage development was organized. The next step was to organize 'train the trainers' programmes for IWASRI and AAPk staff and line agents. Finally, these trainers conducted the training sessions for the farmers, totally 19 sessions, which lasted between 1 to 4 days and were attended by 10 to 75 farmers (depending on the event).

6.1.7   Capacity development for wise use of tropical peatlands

In the 1990s, twenty European and South-East Asian research organizations, universities, NGOs and private consultants joined hands to promote wise use of tropical peatlands in South-east Asia by integrating biophysical, hydrological and socio-economic knowledge [199]. For capacity development, they used the earlier described approach combining the three elements research, education, and advisory services, and initiated projects on this basis. The research projects address issues that are relevant for finding a balance between livelihood (the challenge to increase food production) and resources (the challenge to manage tropical peatlands in a sustainable way). In the education projects, the newly acquired research knowledge is used to develop teaching materials and to implement training programmes. In the advisory projects, the focus is on applying the newly developed knowledge. The partnership varies from project to project depending on the activities and the expertise required for that specific project. The partners are capable in doing this, as they not only work in research, but also in education and training, and in advisory services. As a result, the outputs also cover a wide range of products: from the production of purely explicit knowledge, e.g. in guidelines and handbooks, to building up and using tacit knowledge in joint action/studies and collaborative research activities. The capacity development activities address many aspects as numerous stakeholders, organizations as well as individuals are involved, i.e. (Figure 6.3):

- research organizations and universities: developers of knowledge and users of the end-products;
- international, national and regional government organizations: users of the end-products;
- private companies: co-developers of the knowledge (as they bring in their experiences) and users of the end- products;
- NGOs: co-developers of the knowledge (as they bring in their experiences) and users of the end-products.

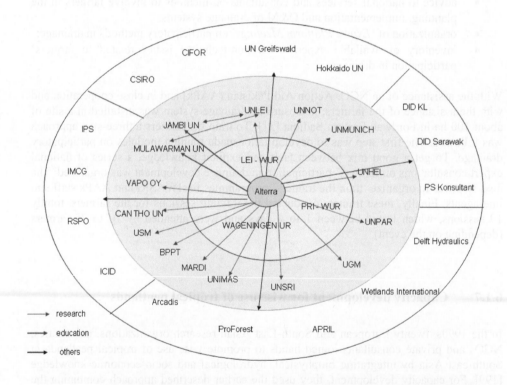

Figure 6.3   In the capacity development projects in SE-Asia many partners and stakeholders are involved: constellation diagram of the Alterra-ILRI partners: (i) inner circle: cooperation within Wageningen University and Research Centre, (ii) middle circle: cooperation with universities and research organizations, and (iii) outer circle: cooperation with the public and private sector (the arrows point to project partnerships).

6.1.8   Discussion

The five examples illustrate an integrated approach for capacity development with the overall aim to improve land and water management by linking tacit and explicit knowledge. If we look to the five strategic steps of capacity development as defined by the WG-CBTE, we see that most activities are related to step 3, 4 and 5 (Figure 6.1). This is not surprising as most activities were undertaken in response to tenders or calls for projects. In most of these projects, except for the recurrent and tailor-made courses and some of the education projects in South-East Asia, the overall objectives were often related to implementation, research and/or advisory services with capacity development as a secondary objective.

The capacity development approach shows great similarity with the knowledge creating process of Nonaka (Figure 6.4). The recurrent and tailor-made courses address the internalization or learning phase. The applied research activities in farmer's fields are the socialization or sharing knowledge phase. The use of this newly developed knowledge to assist local governments with the implementation has characteristics of the externalization or knowledge encoding phase and the dissemination of this new/updated knowledge through publications, papers, course curricula and advisory services are in the combination or synthesis phase. A strict distinction between the activities and the phases is not possible as the knowledge creating process is in principle a never-ending loop. For example, applied research activities in farmer's fields include also elements of internalization, externalization and combination.

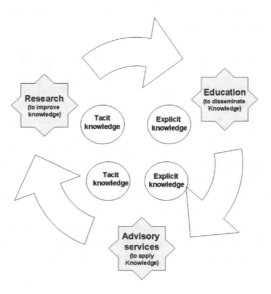

Figure 6.4   Integration of research, education and advisory services: an approach for capacity development in agricultural land drainage based on Nonaka and Takeuchi's knowledge creating process.

There are some essential elements to make this capacity development process successful. The recurrent and tailor-made course programme shows that bringing together professionals from different backgrounds and exposing them to practical knowledge presented from various viewpoints by lecturers from different background/organizations, is an effective tool to integrate explicit and tacit knowledge. A prerequisite is that the participants are stimulated to bring in their own experiences (tacit knowledge) and that the lecturers are capable to link this knowledge to the explicit knowledge they present and their own tacit knowledge. A key element is that most lecturers work part-time in education and part-time in research and/or advisory services. Case studies based on on-going or recently completed projects and studies, in which these lecturers are involved, are an excellent tool to incorporate the latest developments. Refresher courses, in which alumni from the same country or region are brought together to exchange experiences, have elements of internalization and socialization as they link tacit and explicit knowledge. They are an effective tool to keep knowledge up-to-date. That the course programme is valuable can be judged from the fact that many alumni are playing leading roles in the field of drainage and water management [178]. Publishing lecture notes to make the knowledge available to a larger audience has also proven to be an appropriate tool for internalization: the publication series has more than 50 titles and they find their way all over the world.

The capacity development activities in Egypt focused on all four steps in the knowledge creating process: under the guidance of the Advisory Panel, capacity development in research, design and implementation was undertaken. That Egypt has nowadays one of the largest and most modern subsurface drainage programmes in the world can also be attributed to these capacity development activities. This is especially remarkable, because the developments in Egypt took place in a relatively short period: over a period of 40 years manual installation practices were almost completely mechanized, including the introduction of new materials [201].

In India, linking tacit and explicit knowledge was achieved by working together with local researchers and farmers in pilot areas. In this process, the four steps of Nonaka's knowledge creating process were repeated a number of times. The collaboration started with education activities (*internalization*) in combination with joint-field research (*socialization*). The newly acquired knowledge was applied at other universities, in advisory services and to update research and education programmes (*externalization*). Finally, these activities resulted in recommendations to combat waterlogging and salinity for various agro-climatic regions in India (*combination*). To obtain these results, major achievements have been accomplished with human resources development and institutional development. Despite these positive results, major challenges in capacity development remain, in particular to create an enabling environment with the aim to introduce subsurface drainage at a larger scale and, secondly, to improve the performance of water users' organizations with the aim to increase farmers participation [203]. Thus the right mix of tacit and explicit knowledge to improve the externalization activities is still lacking.

The capacity development in Pakistan developed along the same lines and confirms these findings: developing knowledge on technical aspects is easier than initiating participatory drainage development. Nevertheless, even the poor farmers targeted with the participatory drainage work, could manage to contribute about 10% of the total investment cost [21].

The experiences in India and Pakistan confirm the experiences in Egypt, i.e. that capacity development is a long-term process.

Finally, the example from South-East Asia shows that Nonaka's knowledge creating process does not necessarily have to be done in a single project. It can also be achieved on an ad hoc basis in a series of research, education and advisory projects, as long as one or more of the strategic steps defined by the WB-CBTE are addressed and there is a link between the projects, e.g. because the same partners are working together in different projects.

In all examples, capacity development was done in combination with applied research and application. This combination is clearly value for money: it has helped to modernize subsurface drainage practices and considerable savings have been achieved by introducing new methods of investigations, design, planning, installation and O&M [206]. All these innovations could only be implemented because the three elements of capacity development (an enabling environment, institutions and human resource development) were properly addressed and the involved organizations have gone through the knowledge creating process a number of times.

6.1.9   Conclusions

Capacity development in agricultural land drainage is as much a process as a product in which (applied) research, training and education, and advisory services (or extension) are essential elements. The analysis of the examples shows how research, training and advisory services can be linked to the knowledge creating process as described by Nonaka-Takeuchi. By following the four steps of Nonaka's knowledge creating process, explicit knowledge is internalized, e.g. through education and training, and then linked to tacit knowledge by socialization, e.g. applied research. Bringing tacit and explicit knowledge together will yield new knowledge by externalization, e.g. recommendations and advisory services. Thus capacity development is a dynamic process in which the phases of Nonaka's creating process are repeated a number of times. Experience shows that this capacity development process does not necessarily have to be done in one project. It is essential, however, that the three basic elements, i.e. research, education and advisory services, are applied in an integrated manner. This will lead to mutual trust between the cooperating partners and that is much enhanced when there is long-term partnership. It is important that financiers realize that this is a complex and lengthy process that can take many years and that the normally adopted project cycle of 1 to 4 years is generally too short to optimize this process. To make capacity development an effective tool to solve major land and water problems, it is required that financiers, including governments, realize this and that they are willing to invest in such long-term processes. While doing so, the four steps of the knowledge creating process may be useful to design a capacity building strategy.

6.2   Participatory research on the effectiveness of drainage [13]

6.2.1   Introduction

Although the study was conducted in Vietnam, in the humid tropics, the lessons learned with the participatory research are also applicable for research projects in the arid and semi-arid regions, where similar physical and institutional complex situations exist. The Red River Delta (1.7 Mha), located in the north of Vietnam, is one of the most densely populated areas in the world supporting about 1,000 persons per km². An extensive centuries-old system of more than 3,000 km of river dikes and 1,500 km of sea dikes reduces the vulnerability to flooding [165]. Average rainfall varies between 1,600 and 1,800 mm y⁻¹. The rainy season, from May to October, accounts for 80 to 85% of the total yearly rainfall. Rainfall intensities are high, up to 300 to 600 mm in a 3 to 5 day period. Agriculture accounts for about 35% of the gross domestic product, compared to 24% for industry and 41% for services [34]. The Red River Delta is the cradle of the wet rice cultivation in Vietnam, producing about 20% of Vietnam's annual rice production. Rice is planted twice a year and followed by winter crops if possible. Farm holdings are small, on average about 0.3 ha per household. In the Red River Delta, with its low elevations, drainage rather than irrigation is often the limiting factor affecting in agricultural production. The irrigation and drainage systems were designed and constructed in the 1950s and 60s and serve virtually all agricultural land in the Delta. Many of these systems are complex using dual purpose canals and pumped irrigation and drainage. The irrigation and drainage infrastructure was rehabilitated and upgraded under the Red River Delta Water Resources Sector Project (RRWRSP). The project started in 1995 and ended in 2001 [34]. A review of the project showed that irrigation subprojects performed reasonably well, but the two core drainage subprojects performed less than anticipated [29]. The reasons for this inadequate functioning of the drainage systems are diverse and complicated. Firstly, the drainage systems have not been designed and constructed as integrated, comprehensive systems from pumping station (head works) to farmers field, but have gradually expanded over the last 30 - 40 years. Consequently, the capacity of the pumping stations not always matches the capacity of the main canal and field drainage systems [47]. Secondly, given the dynamic situation, the official research and extension system is not always effectively responding to farmers' needs [130]. Thirdly, maintenance, repairs and upgrading practices are poor, resulting in a continuously deterioration of the systems [279]. Fourthly, there is a gradual change in land use: urbanization and non-agricultural use has rapidly increased over the last decades. Water storage in the agricultural fields has also decreased due to changes in cropping pattern, i.e. introduction of high yielding rice varieties and 'dry-foot' crops. These changes have increased the burden on the drainage systems as the non-rice areas have on average less storage capacity and higher runoff intensities. Furthermore, areas capable to hold and regulate excess water such as ponds, lakes, low-lying lands are gradually reclaimed. This has reduced the water storage capacity within the drainage catchment [286]. Finally, the organization of the water management is complicated and fragmentized. The management transfer from government authorities to farmers, initiated in the 1980s, has not yet brought the expected benefits. The management of the drainage system is shared by several organizations, a clear overall responsibility is lacking, Staff is

[13] Published as: Ritzema, H.P., Thinh, L.D., Anh, L.Q., Hanh, D.N., Chien, N.V., Lan, T.N., Kselik, R.A.L., Kim, B.T., 2008. Participatory research on the effectiveness of drainage in the Red River Delta, Vietnam. *Irrigation and Drainage Systems* 22, 19–34. DOI: 10.1007/s10795-007-9028-0

poorly trained and service facilities and funding are insufficient [77; 78]. To overcome these constraints the Second Red River Basin Sector Project (SRRBSP) was initiated in 2002. The SRRBSP promotes integrated water resources management and stakeholder participation at local and basin level. Within the framework of the SRRBSP, a participatory research study was conducted to identify and quantify the major constraints in the functioning of drainage systems and to develop methods to improve the functioning of these systems [278].

6.2.2 Participatory research approach

The hypothesis and scope of the study was that the capacity (physical infrastructure) and operation (institutional infrastructure) of the drainage systems constrains the performance of pumping stations regardless of their discharge capacity. A participatory research approach was adopted to develop conceptual designs to improve the functioning of the drainage systems and to recommend improvements for the institutional capacity of the drainage system management. The research was conducted in two sub-drainage project areas: (i) Phan Dong and (ii) Trieu Duong (Figure 6.5).

Phan Dong sub-drainage area (1956 ha) is located in the Yen Phong district, Bac Ninh Province in the upper reach of the Red River Delta (21^0 13' N – 106^0 04' E). This area has a relatively high elevation (about 2 to 4 m $^+$MSL) and is mainly used for agriculture. Trieu Duong A+B sub-drainage area (4 051 ha) is located in the Tien Lu District, Hung Yen Province in the middle reach of the Red River Delta (20^0 38' N – 106^0 07' E).

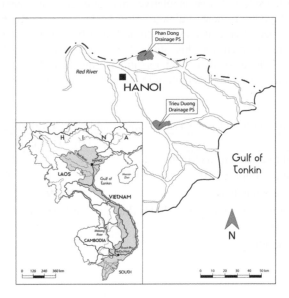

Figure 6.5   Location of the Phan Dong and Trieu Duong study areas

The Trieu Duong area has a lower elevation (1.3 to 3.0 $^+$MSL) with hardly any gradient. Although predominantly used for agriculture, urbanization is increasing rapidly due to the vicinity of Hung Yen town.

The two areas were selected based on a number of criteria; (i) economic re-evaluation, (ii) major constraints in agricultural production, (iii) opinion of local farmers and (iv) size and complexity of the sub-drainage areas (Table 6.2).

Economic re-evaluation
The current internal rates of return (EIRR) are between 4.3 and 7.9%. Much lower than the 12% anticipated during the formulation of the project.

Major constraints in agricultural production
Preliminary appraisal indicates that the capacity of the existing drainage systems (including the pumping stations) is inadequate. Waterlogging and flooding still happens, on average in about 8 to 12% of the cropped areas, resulting in reduced agricultural productivity.

Opinion of local farmers
The farmers in the study areas are not really satisfied with the project effectiveness because, since the upgrading of the Trieu Duong and Phan Dong pumping stations, partial water-logging and flooding still happen. Yields in the affected areas are well below the average yield.

Size and complexity of the sub-drainage areas
Although both areas are representative for the prevailing conditions in the Red River Delta polders, the size of the selected areas is relative small and their catchments are more or less independent. The layout of most drainage systems in the Red River polders is very complex: they have been frequently expanded and modified over the last thirty year. The two areas that were selected for this study have relatively straightforward boundary conditions. This was done to avoid that too much time and effort had to be spent on understanding the often complex interaction in the water management between polders.

Table 6.2   Selection criteria for the Trieu Duong and Phan Dong sub-drainage areas

Name of project	Trieu Duong B	Phan Dong
Economic internal rates of return (EIRR in %):		
- Project Completion Report [29]	6.1	4.3
- Calculated by the Study team	7.9	7.3
Area prone to flooding (average 1997 – 2004) (ha)	329 (8%)	234 (12%)
Summer rice yield (t/ha):		
- Area prone to flooding	4.6 - 5.1	3.3 - 4.6
- Total area	5.0 - 6.1	3.7 - 5.2

Participatory learning and action

The project adopted the Participatory Learning and Action (PLA) approach [87]. The step-wise approach is presented in Figure 6.6. To ensure stakeholder participation, several PLA workshops were organized. At the start of the project, to establish sub-project drainage committees and to quantify the stakeholder's views on the problems, constraints and improvement options. And, at the end of the project, to review and prioritize the conceptual design options to improve the functioning of the drainage systems.

The outcomes of the initial workshops were used to develop and conduct participatory pre-investigation programmes. The following data were collected: catchment area boundaries, topography, land use, design criteria and layouts of irrigation and drainage systems, social-economic and environmental data. To understand the institutional set-up, additional data were collected on the organization of the water management, including funding and O&M practices. For the participatory pre-investigations the following tools were used: village profile, village diagram, cropping calendar, economic classification at household level and Venn diagram [87]. These tools were used to identify the stakeholders and, together with the stakeholders, to analyse and prioritize the problems (problem tree) and to formulate and recommend improvement options.

The findings of the pre-drainage investigations were used to develop and implement a monitoring programme during the rainy season of 2005 (May to October). The main objective of the monitoring programme was to collect sufficient data to model the drainage canal systems. The following data was collected: rainfall, evaporation, water levels in rice fields and in the main and secondary canal systems, pumping hours of the main pumping stations, operation practices of the pumping stations and control structures in the canal system, water quality parameters, land use and crop yield, and cost of production.

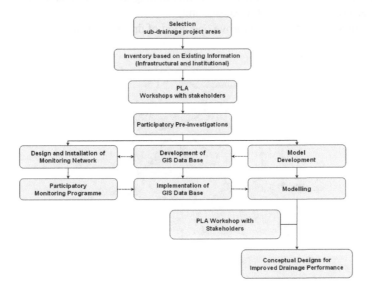

Figure 6.6   The participatory research approach adopted by the study team

The data were stored in a GIS data base and were used to model the drainage systems. The systems were modelled with 'DUFLOW', a one-dimensional, non-steady state, model for water movement and water quality [147]. DUFLOW was used to simulate the hydraulic functioning of the drainage system. In particular the complex relation and interaction between the various elements of the system, i.e. drainage canal sections, structures connecting these sections (culverts, siphons, etc), regulation structures (gates, etc) and the pumping station(s) (Figure 6.7). After calibration, a number of options to improve the functioning of the drainage systems were simulated. These options included, among others:

- installation of trash racks, not only at the intake of the pumping stations, but also at specific locations in the main drainage system;
- remodelling the main drains;
- installation of more culverts and regulators, and
- alternatives for the operation of the pumping stations and regulators.

In PLA workshops, these improvement options were discussed with the stakeholders. Based on their recommendations, an implementation manual for the planning and design of future drainage upgrading projects was prepared.

6.2.3   Pre-drainage investigations

Four PLA Workshops were organized:
- to establish the Sub-project Drainage Committees (SDC);
- to collect information on irrigation and drainage practices;
- to identify the exact boundaries of the drainage areas;
- to identify the constraints in the functioning of the drainage systems and the stakeholders views on the causes, and;
- to propose solutions to improve operation of drainage systems.

Figure 6.7   The lay-out of drainage systems in the Red River Delta is complicated (Example Trieu Duong area).

Technical as well as institutional (non-technical in terms society and management) aspects were taken into consideration. The participants of the workshop were representatives (both male and female) of farmers, communes, local government, unions, NGOs, etc. They were selected from:

- provincial department of agriculture and rural development (DARD) and people's committees;
- irrigation and drainage management committees (IDMC) and hydraulic groups;
- at commune level: leaders of the people's committees and women and farmers' associations;
- at village level: head of villages, irrigation teams and farmers.

In the workshops, the stakeholders identified their priorities and their preferences:

- to improve the functioning of the drainage system and to reduce risks of water logging and flooding;
- to increase productivity of crops;
- to improve economy within the framework of the policy and overall development plans of the local government.

The workshops were also used to determine the various responsibilities of all stakeholders (groups). Furthermore, stakeholders were mobilized to participate in the monitoring programme. The workshops proved to be useful to help the stakeholders, especially the farmers, to be more confident in identifying, analyzing and recommending proposals. All participants agreed on the need to establish SDCs. The functions and tasks of the SDCs, including the required qualifications to be eligible, the membership ratio male/female, etc., were agreed upon and lay down in regulations. SDCs were established and members elected. The members are experienced farmers, both male and female. They have prestige, responsibility and show willingness to represent the farmers of the project areas. The stakeholders assessed and ranked the problems they encounter with draining their fields (Table 6.3). The ranking shows that, next to technical constraints in the infrastructure, the institutional constraints are equally recognized. It is of interest to note that farmers not only blame the authorities but also realize that their own attitude can be improved.

6.2.4   Monitoring programme

Through the monitoring programme the drainage problems were quantified. The problems range from the functioning of the main pumping stations, the main drainage system up to the tertiary and field level.

Actual pumping capacity
The actual discharge capacity of the pumping stations is less than the design capacities; respectively 92 and 80% for Phan Dong and Trieu Duong (Table 6.4).

Table 6.3   Ranking of the problems encountered in drainage as assessed by the stakeholders in the two sub-drainage areas

Rank no.	Trieu Duong Problem	Rank no.	Phan Dong Problem
1	Lack of culvert gates and valves	1	Some drains are too small
2	Lack of regulators	2	Inadequate regulations for violation(s)
3	IDMC is not active due to lack of funding	3	Regulators are operated improperly.
4	It is not possible to regulate water levels in sub-areas	4	Budget to dredge the drainage canals is insufficient.
5	Drainage outlets to the rivers are too small	5	Operation in (some) sub-areas hampers the functioning of the main system.
6	Investments are not done systematically	6	Monitoring by local authorities is not in time.
7	Instructions from leadership are inadequate	7	Monitoring of IDMC is not in time and careless.
8	Awareness of farmers is insufficient	8	Supervision, assessment, reports are unrealistic.
9		9	Awareness of farmer is limited.
10		10	Propaganda on canal protection is limited / has constraints
11		11	Pumping station is the main source of the problems.

Table 6.4   Results of the discharge measurements at Trieu Duong Pumping Station (each pump was calibrated twice)

Pump	Calibration No. 1			Calibration no. 2		
	Measured discharge (Q_m)	Design discharge (Q_d)	Ratio Q_m/Q_d	Measured discharge (Q_m)	Design discharge (Q_d)	Ratio Q_m/Q_d
	$(m^3 s^{-1})$	$(m^3 s^{-1})$	(-)	$(m^3 s^{-1})$	$(m^3 s^{-1})$	(-)
No.1	2.05	2.39	85.8	1.83	2.39	76.6
No.2	2.09	2.37	88.2	1.86	2.38	78.2
No.3	2.00	2.39	83.7	1.73	2.37	73.0
No.4	2.04	2.39	85.4	1.79	2.36	75.8
No.5	2.03	2.39	84.9	1.76	2.36	74.6
Average	2.04	2.39	85.6	1.79	2.37	75.6

Design pumping capacity

The design capacity of the pumping stations had been underestimated because the design is based on the drainage requirements for rice crops only. In reality, only 52 to 68 % of the land is used for rice cultivation (Table 6.5).

Table 6.5   Change in land use in Phan Dong and Trieu Duong areas

	Land use							
	Phan Dong				Trieu Duong			
	1996		2005		1996		2003	
	(ha)	(%)	(ha)	(%)	(ha)	(%)	(ha)	(%)
Agricultural use:								
Rice-double cropping	1.088	68	1.281	66	2.137	54	2.129	53
Upland crops	59	4	81	4		0		0
Aquaculture	0	0	0	0	215	5	235	6
Others	36	2	30	2	256	6	289	7
Sub-total	1.183	74	1.392	71	2.608	66	2.652	65
Non-agriculture use:								
Villages and farm building	83	5	160	8	305	8	358	9
Grave yard	18	1	18	1	45	1	46	1
Borrow bits (for brick making)	11	1	48	2	16	0	12	0
Road	92	6	209	11	175	4	228	6
Others (including waste land)	220	14	130	7	807	20	756	19
Sub-total	424	26	563	29	1.347	34	1.399	35
Total	1.607	100	1.956	100	3.955	100	4.051	100

The rest of the land is used for the cultivation of other crops (mainly maize, vegetables and tree crops) and non-agricultural uses (i.e. roads, villages, town etc). These none-rice uses have significant lower in-field storage capacity. Furthermore, land use has changed over the last 10 years. These land use changes require a discharge capacity that is higher than the design capacity: 12 to 18% for respectively Phan Dong and Trieu-Duong.

Effective pumping time
The effective pumping time is less than expected because the suction basins of the pumping stations are frequently blocked by floating debris. During periods of peak drainage demand, pumping has to be stopped for 2 to 3 hours per day to remove this debris. The poor functioning of the pumping stations results in higher water levels in the sub-drainage areas and increases the risk of flooding. This risk of flooding is even higher because the functioning of the main drainage system is also below expectations.

Main drainage system: bed levels and cross-sections
Main and secondary drains have higher bed levels and wider cross-sections than the design values. Fortunately, these two effects more or less neutralize each other. Thus, the hydraulic capacity of the main drainage system is in general in balance with the capacity of the pumping stations. The overall capacity of the drainage canals, however, is below the design capacity. At various locations, the canal cross-sections are too small, especially where the drainage canals pass through villages.

Structures

Culverts and regulation structures are either too small or frequently blocked (without authorization) by individual farmers or group of farmers, e.g. to irrigate fields with higher elevations or to storage water for aquaculture. This results in an increase in upstream water levels.

(Mis-)Use of canal sections.

Water levels in the drainage canal systems are also higher because some canal sections are (mis-)used for fishing (nets are installed across the drains) and used to dump farm and other waste products. The resulting huge amounts of floating debris restrict the flow and block culverts or other structures.

Tertiary and on-field drainage systems

The capacity of the tertiary and on-field drainage systems is poor because these systems were not properly designed and/or constructed. Most on-farm outlets and field drains are missing, broken or damaged. At present, it is hardly possible to regulate the water levels in the fields and/or the discharge to the main drainage system.

Water quality

The water is not only polluted by floating debris, but also by domestic waste water and waste water from (small) industrial enterprises. To assess the water quality the following parameters were monitored: temperature, turbidity, pH, nitrite (NO_3), phosphate (PO_4), bio-oxygen demand (BOD), total solids (TS), dissolved oxygen (DO) and faecal coli form. The water quality was classified from 0 (heavy polluted) to 100 (very good water quality) using a method developed by Canter [46]. The water in the sub-drainage areas is not seriously polluted, although the quality is slightly lower than the quality in the adjacent rivers (Figure 6.8). The variation between the dry and wet season is more pronounced. In the months before the rains start (April – June), the water is more polluted than during and shortly after the rainy season (September-December).

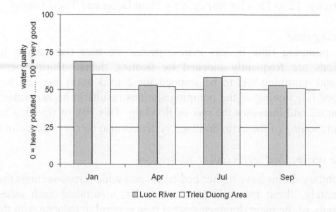

Figure 6.8   Water quality in the Trieu Duong sub-drainage area and in the adjacent Luoc River. Water quality is classified from 0-25: heavy polluted; 25-50: polluted; 50-70: normal; 70-90: good; 90-100: very good (after Canter [46]).

Next to the infrastructural constraints as discussed above, there are numerous institutional constraints that hamper the functioning of the drainage system, i.e.:

- *complex ownership*: the boundaries of the sub-drainage areas and the underlying sub-division in secondary and tertiary units do not coincide with the boundaries of villages, communes. This complicates the organization of the water management.
- *many organizations*: many organizations are involved (Figure 6.9) and their responsibilities are not always clear, transparent and specific (Figure 6.10).
- *poor coordination*: the coordination between these organizations is not adequate. The Government does not put much emphasis on a participatory approach in O&M. Policy mechanism and specific guidance on participation are lacking. As a consequence the farmers look after their own benefits and not to the benefits of the commune, village or sub-drainage area.
- *unbalanced investments*: the Government only invests in the head works (main pumping stations) and primary canals. Farmers are responsible for the tertiary and on-farm infrastructure.
- *lack of funds*: drainage rates are included in the water fees. These fees, however, are insufficient: at present water fees cover only about 10-18% of the O&M costs.
- *inadequate monitoring network*: an adequate monitoring system is lacking. Only a few staff gauges have been installed and are monitored, mainly at the main pumping stations.
- *Limited capacity to control drainage*: there are not enough structures to regulate the flow in the drainage systems, and if there are structures the dimensions and levels are often incorrect. Furthermore, water regulation is based on local-specific preferences often obstructing upstream water management practices.
- *low confidence*: the majority of the farmers and households lack confidence in drainage management (Table 6.3).

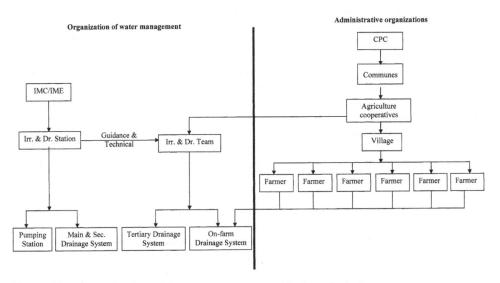

Figure 6.9   Organization of the water management in the sub-drainage areas

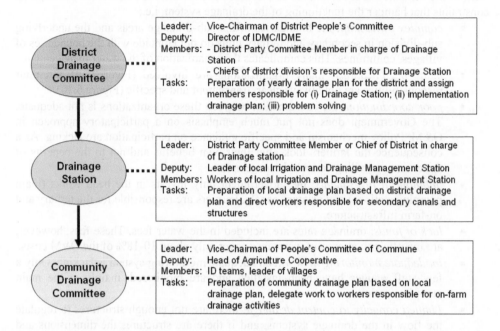

Figure 6.10   Task Forces for Drainage Control to be established each year prior to the
rainy season (B)

The infrastructural and non-infrastructural constraints in the functioning of the drainage systems have major repercussions. Every year, parts of the areas suffer from waterlogging and flooding, on average 12 and 8% of respectively Phan Dong and Trieu Duong. This flooding is not so much related to the topography but more to:

- *location*: in rice fields adjacent to the main drains, the depth of the standing water is always lower than 6.0 cm, even after heavy rainfall (Figure 6.11). This is well inside the safe limits for rice cultivation. However, in fields further away from the drains or in upstream sections, the depth of the standing water can rise to 45 cm, well above the maximum allowable level. This indicates that the field and tertiary drainage systems are inadequate.
- *land use*: flooding occurs upstream of fields used for aqua-culture and canal sections where farmers block drains to irrigate lands with slightly higher elevation.

As a result of the poor functioning of the drainage systems, rice yields in the inundation-prone areas are 10 to 14% (Phan Dong) and 8 to 23% (Trieu Duong) below the overall average rice yields in the areas. Thus, it is not surprising that the actual economic internal rates of return, i.e. 7.3% for Phan Dong and 7.9% for Trieu Duong, are well below the anticipated 12%.

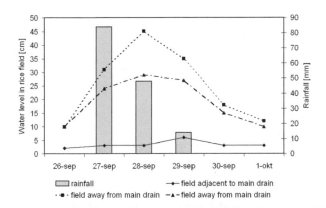

Figure 6.11   Water levels in rice fields near the main drains are lower compared to water
levels in fields away from the drain (Data from Phan Dong area)

6.2.4   Model simulations

The DUFLOW model was calibrated with the data collected during the participatory
monitoring programme. The calibration was done as follows: actual dimensions of the
drains and related structures were matched with the measured water levels at various
locations in the drainage system during days with heavy rainfall and the corresponding
discharges of the pumping stations (Figure 6.12). The resistance in the drainage system
(roughness) was used to match the measured and simulated water levels. After calibration,
the model was used to simulate:

- *design situation*: to verify whether the design capacity of the pumping stations is
  in line with the design capacity of the canal system;
- *actual situation*: to assess the capacities of the existing canal sections and
  associated structures and to check whether the installed pumping capacity is
  sufficient during 'normal' operation conditions (Figure 6.13);
- *extreme conditions*: to assess the functioning of the system during extreme rainfall
  events recorded over the last 5 to 10 year;
- *improvement options*: to get a better match between the capacity of the drainage
  canals (with related structures) and the pumping station.

6.2.5   Conclusion and recommendations

Under the first RRDWSP, the increase in agricultural production in most of the project
area was less than anticipated. To assess the performance of the sub-drainage projects, a
participatory research study was conducted in two areas. The hypothesis and scope of the
study was that the capacity (physical infrastructure) and operation (institutional
infrastructure) of drainage system constrains the performance of pumping stations
regardless of their evacuation capacity. A participatory monitoring programme revealed
that this hypothesis is partly correct.

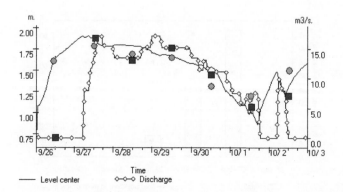

Figure 6.12   Example of the model calibration: measured (dots) and simulated (lines) upstream water levels and discharges at Trieu Duong B pumping station.

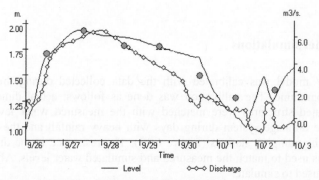

Figure 6.13   Measured (dots) and simulated (lines) upstream water levels and discharge through a road culvert in Canal T1 in Trieu Duong sub-drainage area

The mainly reasons are:
- the installed pumping capacity is only 80 to 92% of the capacity specified in the design;
- the design capacity has been underestimated because it was based on the drainage requirements for rice only and in reality about one-third (Phan Dong) to almost 50% (Trieu Duong) of the land is used for other crops and non-agricultural purposes. These non-rice land uses required a higher drainage capacity, and;
- the effective pumping time is reduced because during extreme rainfall events the suction basins of the pumping stations or blocked by floating debris. On average, pumping has to be stopped 2 to 3 hours per day to remove the debris.

Based on the results of the monitoring program and the computer simulations carried out with the DUFLOW Water Flow model, a large number of options to improve the functioning of the drainage systems were developed, i.e.:
- the capacity of the pumping stations can be increased by the installation of trash racks, equipped with automatic debris removal devices. This measure will

increase the total pumping time by 2 to 3 hours per day. It will also reduce the head loss over the pumping station and thus increase the discharge capacity.

- the drainage design rate should not be based on rice cultivation only, but should take into account the percentage of land used for non-rice crops (maize, vegetables and fruit trees) and non-agricultural use (villages, town, road, grave yards, etc.).
- the functioning of the main drainage system can be improved. At present, structures are often too small compared to the capacity of the main canals. Furthermore, these structures should be designed and operated as control structures. This will allow farmers or farmers groups to irrigate fields with higher elevations or to store water for aquaculture without hampering the functioning of the drainage in the upstream areas. Trash racks should be installed to avoid that rubbish or debris ends up in the downstream parts of the system.
- the tertiary and on-farm drainage systems need to be designed and implemented based on the drainage requirements for the various types of land use, i.e. agriculture (rice crops, vegetables, fruit trees, etc), aquaculture, etc.

During the final PLA workshops, the stakeholders prioritized the improvement options. The study team estimates that, if these prioritized improvement options are implemented the EIRR will increase from 7.9 to 14.5 % and from 7.3 to 13.0% for respectively Trieu Duong and Phan Dong. It should be realized, however, that this increase is not only the result of improved drainage, but also of land use changes and adjusting input data such as O& M cost, actual rice periodicity and market prices. Especially the land use changes have a major impact as some low-lying areas cannot be drained economically: 329 ha and 232 ha in respectively Trieu Duong and Phan Dong. In the simulations, these low-lying areas have been converted from rice into aquaculture. Furthermore it should be realized that, to benefit from these improvement options, the institutional set-up also has to be improved. There are various options to do this. It is recommended to use the following guidelines:

- *clear and transparent responsibilities*: give one organization the responsibility of a well-defined part of the drainage system and avoid overlap between organizations.
- *responsibility at the lowest possible level*: at farm level, the farmers, being the main stakeholders, have to be made responsible for the on-farm drainage system and management. At tertiary level, the farmer's organizations (such as water-use cooperatives, water-use groups, agricultural cooperatives serving in irrigation and drainage) should be responsible.
- *include all stakeholders*: the drainage system not only serves agricultural lands, but also villages, sometimes even small towns, industrial sites, etc. All these non-agricultural stakeholders or users should be included.
- *charge all stakeholders*: water fees should not only be collected from the farmers, but all stakeholders should pay based on the benefits they receive.
- *need for monitoring*: for proper operation of the pumping stations and the control structures in the drainage system, an adequate monitoring system is required. Water level gauges at various locations have to be installed and monitored.
- *need for capacity building*: drainage management in these flat polder areas is complicated and management practices at various levels are very much interrelated. All stakeholders, from the individual farmer to the operator of the

pumping station, need to be properly trained in the complex drainage management practices.

To apply the results of this participatory research study in other areas in the Red River Delta, the Study Team has prepared an 'Implementation Manual' [277]. The manual presents a method for a participatory diagnostic process to identify and to qualify constraints in the functioning of drainage systems. Next, the manual can be used to prepare conceptual design options to improve the functioning of these systems. Technical and non-technical (institutional) improvement measures can be developed based on the prevailing socio-economic and environmental conditions. Based on a PLA approach, stakeholders can discuss, select and agree upon measures to improve the functioning of the drainage systems. For the development of these improvement measures the following information is essential:

- flooding conditions and topography corresponding to different land-use scenarios;
- drainage effectiveness in relation with changes in cropping patterns and changes in land use.

The study has shown that a participatory approach, including representatives of all stakeholders, is an effective and efficient method to assess the effectiveness of drainage in the Red River Delta. These stakeholders included not only the farmers, but also the other people living the densely populated delta and their organizations. The study confirmed that the capacity (physical infrastructure) and operation (institutional infrastructure) of the drainage systems indeed constrains the performance of pumping stations. Only close cooperation between all stakeholders can improve the drainage in such complex systems as the polders in the Red River Delta. Furthermore, a combination of technical and institutional measures is required to improve the functioning of the drainage systems.

7   Synthesis: subsurface drainage practices in irrigated agriculture

7.1   Is subsurface drainage an acceptable option?

7.1.1   Are subsurface drainage systems technically sound?

Subsurface drainage has been practised for thousands of years, but its use on a large scale only started around the middle of the last century when the prevailing empirical knowledge of drainage and salinity control was given a solid theoretical foundation [41]. Over the past 25 years, the role of drainage has changed from a single-purpose measure for controlling waterlogging and/or salinity to an essential element of integrated water management under multiple land use [219]. Most theories were developed for subsurface drainage in the temperate zone, mainly in Western Europe and the USA [192; 228]. Although these theories now form the basis of modern drainage systems, there will always remain an element of art in land drainage [41]. A desalinization study conducted in one of the subsurface drainage pilot areas in India shows that empirical equations have to be adapted to fit local circumstances [242]. It is not possible to give beforehand a clear-cut theoretical solution for each and every drainage problem: sound engineering judgement on the spot is still needed, and will remain so.

The subsurface drainage systems that are installed in Egypt, India and Pakistan show great similarities; they are composite systems consisting of buried collector and field drains. This is quite remarkable, because, although in the three countries irrigated agriculture is practised by small marginal farmers with landholdings of only a few hectares, the soil, hydrological, climate and socioeconomic conditions vary considerable. As rainfall levels in Egypt are negligible, agriculture depends almost entirely on irrigation from the River Nile. The soils in the Nile Valley and Delta, the major agricultural areas, are rather uniform, consisting mostly of light to heavy clays, with a clay content ranging from 30 to 80% [23]. Environmental conditions in India and Pakistan are much more diverse: agriculture depends also on the monsoon rainfall, which varies from North to South and East to West, soil texture ranges from coarse sandy soils to heavy black cotton soils, and the social setting is much more diverse and complex. In Egypt, the collectors discharge by gravity into the main open drainage system, from where the drainage effluent is either drained into the River Nile, pumped back into the irrigation system or discharged to the Mediterranean Sea [149]. In Pakistan, the collectors discharge into a sump from where the drainage effluent is pumped into the open main drainage network, from where it is discharged back to the Indus River or through the left-bank outfall drain to the Arabian Sea [301]. In India, both gravity and pumped systems are used, mostly draining back to the major rivers [150].

In all three countries, the subsurface drainage systems are designed using steady-state equations. Although irrigated agriculture results in considerable water table fluctuations during the growing season, the steady-state approach can be used because the change of storage over the season is small compared with the volume of recharge and discharge [156]. The depth and spacing of the field drains differ between the countries (Table 7.1).

Table 7.1   Characteristics of the subsurface drainage systems used in Egypt, India and
             Pakistan [150]

	Egypt	India	Pakistan
Size of unit (ha)	< 120	50	100–450
Design:			
- discharge (mm d^{-1})	1.0	1.5–2.0	0.95–3.5
- water table (m)	1.0	> 1.75	1.0–1.2
Drain depth:			
- field drains (m)	1.2–1.5	1.0–1.75	1.8–2.4
- collector drains (m)	< 2.5	< 3.0	< 3.0
Drain spacing (m)	30–60	45–150	60–300
Field drains:			
- material	PVC	Concrete & PVC	PVC
- diameter (mm)	100	100	100–200
- length (m)	200	< 315 m	800
Collector drains:			
- material	PVC or HDPE	PVC	PVC & PE
- diameter (mm)	200–400	80–450	200–380
- length (km)	< 5	< 1	< 4
Envelopes:			
- material	gravel & synthetic	gravel & synthetic	gravel
- used in soils with			
clay content	< 30%	30–40%	

In Egypt, relative shallow systems have been installed, with drain depths up to 1.50 m. In Pakistan and India, drains are installed at greater depths (> 1.75 m), based on the critical depth concept, to avoid secondary salinization caused by the upward flux of water once the water table rises to 2–3 m below the soil surface [91]. The research conducted in the farmers' controlled pilot areas in the Nile Delta in Egypt and in five agroclimatic regions in India show that drain depths and design discharges can be reduced. Research conducted in Pakistan confirms these findings. In Egypt, a design discharge rate of 0.9 mm d^{-1} is sufficient to cope with the prevailing losses of irrigation water and to maintain favourable soil salinity levels [8]. This is 10% lower than assumed in the design. Subsequently, the design discharge rates for collector drains can also be decreased [69]. In India, research shows that the original design rate for salinity control (2.0 mm d^{-1}) can be reduced to 1.0–1.5 mm d^{-1} [203; 243]. In Pakistan, field monitoring programmes and computer simulations indicate that the field drainage design discharge can be reduced from the initial value of 3.5 mm d^{-1} to 1.5 mm d^{-1} [291]. The same applies to the depth of the drains. In Egypt, pilot area research showed that a design depth for the water table of 0.80 m proved to be sufficient [8]. The most cost-effective way to obtain this depth at the given discharge is to install drains at a depth of between 1.20 to 1.40 m [149].

In India, field data combined with simulations using SALTMOD indicate that drain depth, under gravity flow conditions, can be reduced to 0.9–1.0 m [203; 244]. Research conducted in the Tungabhadra irrigation project in Karnataka, India, shows that, alternatively, drain spacing can be enlarged [136]. In Pakistan, the design depth has also been gradually reduced from a drain depth of 2.25–2.40 m in the 1980s to 1.50–2.10 m in the 1990s [177]. These research results support the widely prevailing view that deep drains are unnecessary for salinity control in irrigated lands [234]. Reducing drain depth does not generally deliver major savings in the cost of implementation, the savings are more environmental. Salt from the deeper subsoil is not disturbed, resulting in a lower salt load in the drainage effluent. A simulation study for a sandy loam soil in North Carolina, USA, indicates that shallower drains also reduce nitrogen losses from subsurface drainage [227]. Shallower drainage systems are therefore likely to have less environmental impact on the downstream areas. The introduction of operational control, required to get a better integration between irrigation and drainage, can further reduce drainage intensities.

Over the years, design practices have been modified, not only to incorporate the changes in design discharges and drain depth but also to optimize subsurface drainage for areas where crops with different crop water requirements are cultivated [206]. The introduction of simulation models in the 1980s has greatly improved knowledge of the functioning of subsurface drainage, the design of these systems and analyses of water and salt movement under varying and complex field conditions, as in the case of integrated irrigation and drainage management [5].

Installation practices have evolved from purely manual installation on individual farm plots to fully mechanized installation programmes covering thousands of hectares [201]. To make this rapid change possible, practical tools for implementation had to be developed, starting with the introduction of new types of installation equipment, trencher and trenchless drainage machines. To optimize the use of these machines a number of problems had to be solved. New materials for drain pipes and envelopes had to be developed to reduce the high transportation and installation cost of the traditional materials and to improve the quality of construction. New methods for quality control had to be developed as the traditional methods proved to be inadequate because of the increased speed and method of mechanical installation. Computerization was introduced to increase the pace of implementation and reduce costs, especially of large-scale projects. Moreover, staff had to be trained in these modernized drainage machinery and installation techniques, as well as in the planning and organization of the implementation process.

The case studies presented in Chapters 2, 3 and 4 show that the subsurface drainage systems installed in Egypt, India and Pakistan effectively prevent waterlogging and root zone salinity in irrigated lands. The case studies show that subsurface drainage systems have not only reduced the risk of waterlogging and salinity but also enhanced crop production because they enable supplementary measures in soil and water management, like the application of gypsum, the introduction of salt-tolerant varieties and irrigation efficiency improvement.

The water and salt balance studies conducted in Egypt and India show that rice cultivation plays an important role in the leaching of salts from the soil profile, reducing the drainage intensity during the cultivation of other crops.

There is abundant evidence that these subsurface drainage systems increase yields and therefore rural incomes. In Egypt, a nationwide monitoring programme revealed the following improvements [20]:

- average water tables (5 days after irrigation) significantly decreased from about 0.6 m before drainage to about 0.9 m four years after the installation of subsurface drainage.
- areas with saline soils decreased from 80% (before drainage) to 30% (after drainage) in saline areas and from 40% (before) to 5% (after) in non-saline areas.
- yields of all crops increased, possibly more than expected, although individual crops reacted differently. This is in agreement with the research conducted in the Mashtul pilot area where the increase was 10% for rice, 48% for *berseem*, 75% for maize and more than 130% for wheat. These yield increases can be attributed in part to the decrease in soil salinity and in part to the effect of improved water and air conditions in the root zone and improved agricultural inputs [14].

Similar results were obtained for the subsurface drainage projects in India. In the five agroclimatic regions where canal irrigation is most important, crop yields increased significantly: on average 54% for sugarcane, 64% for cotton, 69% for rice and 136% for wheat [203]. These yield increases were obtained because in the drained fields water tables and soil salinity levels were respectively 25% and 50% lower than in the non-drained fields. Similar results were recorded in the large-scale projects: in the RAJAD Project (15,000 ha) the overall crop yield increased by about 25% [150]. In Haryana, the effects of subsurface drainage in an area of 1,200 ha were compared with a non-drained area of 1,000 ha [222]. In the drained area, the increase in the yields of different crops ranged from 19 to 28% compared to a decrease in the non-drained area due to increased waterlogging and soil salinity problems. In the drained area, the average soil salinity decreased by 36% and the water table was deeper, especially during critical periods (monsoon season).

In Pakistan, subsurface drainage systems control the water table, decrease soil salinity, increase crop yields and crop intensity and decrease the area of abandoned lands [38]. For example, the waterlogged conditions in Mardan SCARP reduced considerably after the project was completed: a monitoring programme revealed that the post-drainage water table fluctuates between 1.2 and 2.5 m below ground level, compared to a pre-drainage fluctuation of between 0.3 and 1.2 m [150]. The subsurface drainage systems, although more expensive, are better for the environment than tubewell drainage systems. After the installation of subsurface drainage, the shallow groundwater quality improved, whereas the deep groundwater quality did not change [294]. In contrast, in the areas drained by tubewells the groundwater quality remained constant or even deteriorated. This confirms the conclusion that shallower, less intensive drainage systems cause lower environmental impacts.

It can be concluded that the subsurface drainage systems installed in Egypt, India and Pakistan effectively prevent waterlogging and root zone salinity in irrigated lands and consequently increase crop yields and rural income. The research supports the prevailing view that deep drains are unnecessary for salinity control and that improved operational management can further reduce drain depths and design discharges. The introduction of new types of installation equipment and materials and the corresponding implementation practices has made large-scale implementation feasible.

7.1.2   Are the subsurface drainage systems cost-effective?

The costs of installing large-scale subsurface drainage systems depend on local physical and economic conditions and the installation method (Table 5.1). As a result the overall cost varies considerable. In Egypt the overall cost per hectare is € 750 [20], in India the cost per hectare in large-scale schemes varied from € 770 for the HOPP Project to € 815 for the RAJAD Project [164] and in Pakistan the cost for the EKTD Project was € 1,200 per hectare [150]. These investments proved to be very cost-effective.

In Egypt, the Gross Production Values increased with € 500–550 per hectare and the annual net farm income of the traditional farm increased by € 375 per hectare in non-saline areas and by € 200 per hectare in saline areas [20]. The payback period was no more than 3 to 4 years. The impact of drainage on national agricultural production is also significant; drainage accounted for about 8% of production in the agricultural sector. The contribution to the gross domestic product is estimated at about € 0.9 billion per year. The government prefinances the total cost of the installation of subsurface drainage. Farmers payback these costs over 20 years with a grace period of 3 to 4 years without interest, which effectively amounts to roughly a 50% subsidy [7]. The land tax is also slightly increased to pay for maintenance.

Similar results have been reported in India: benefit-cost ratios vary from 1.2 to 3.2, internal rates of return from 20 to 58% and payback periods from 3 to 9 years (Table 3.7). The value of the land increased as well: in Gujarat subsurface drainage pilot areas attracted farmers to buy land in the area at prices up to 5 times higher than the pre-drainage period. These results, obtained in farmers' fields but on a small-scale, are in agreement with results found in Rajasthan. The installation of subsurface drainage in the RAJAD Project (15,000 ha) proved to be economically sound: the benefit-cost ratio ranged from 1.3 to 2.9, the net present value ranged from € 200 to € 1,050 per hectare, the payback period was from 4 to 7 years and the internal rate of return ranged from 18 to 35% [86]. In Haryana, land use intensified after installation of subsurface drainage, cropping patterns changed in favour of more remunerative crops and crop yields increased [56]. Drainage helped to increase land productivity, gainful employment of farmers and farm income. The financial and economic feasibility of drainage in waterlogged and saline areas looks favourable, provided sufficient water is available for leaching and irrigation and that a sustainable solution for the disposal of the low-quality drainage effluent is found. In India, funding of about € 635 per hectare is provided by the central government, the state governments and farmers in the ratio of 50 : 40 : 10 [86]. Farmers, both male and female, are willing to pay their part of the cost as they clearly see the benefits of drainage [203]. In reality, however, they are too poor to pay their part of the installation cost.

In Pakistan, installation costs are heavily subsidized by the provision of low cost loans, low cost renting of equipment and free technical assistance [37]. While the potential of pricing and subsidizing as instruments for waterlogging and salinity control is recognized, implementation has been found to be complex. The drainage fees paid by the farmers are not even enough to pay for O&M: they cover only around 20% of the actual expenses [150].

The case study on a modified layout shows that for a slightly higher cost (between 6 and 12%) up to 30% savings in irrigation water supply and up to 25% in reduced drain discharge can be achieved.

It can be concluded that subsurface drainage systems are a very cost-effective measure for combating waterlogging and salinity in irrigated agriculture. The recent rise in the price of major food commodity prices (Figure 7.1) will increase the economic returns even further. It should be remembered, however, that most farmers are poor and do not have the means to invest in subsurface drainage. A farmer in South Punjab, Pakistan, who was asked *'When would you consider yourself to be a rich farmer?'* answered *'When I can give my family three meals a day'* (personal communication Knops, 1999).

7.1.3   Is subsurface drainage a socially accepted practice?

In Egypt, the drainage programme is implemented by the EPADP. During the last two decades, participation by farmers has been gradually introduced, mainly in O&M [7]. Like the Water Users Associations (WUA) for irrigation, Collector User Groups (CUG) were established which are responsible for cleaning manholes and undertaking minor repairs. They were not very effective because the farmers did not feel sufficiently motivated to do collective work. Furthermore, there is the problem of discrepancy between the responsibility of the WUAs (areas served by tertiary units) and the CUGs (areas served by a collector). To solve these problems, farmers' participation has been scaled up to district level and Water Boards have been established. Water Boards are formed by members of WUAs and households to integrate water supply and sanitation and to represent all users' interests in water management. A problem, however, is that the Water Boards do not have the legal powers to collect fees and the MWRI is not allowed to subsidize the Water Boards [289]. At agency level (MWRI) the local irrigation districts and centres have been combined into Integrated Water Management Districts to provide advisory services and monitoring activities.

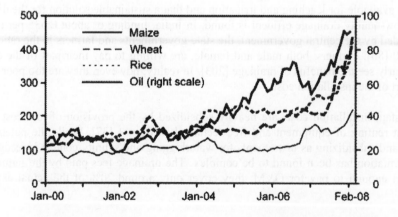

Figure 7.1   World Commodity Prices, January 2000 to February 2008 (US$/tonne) [283]

In India, there is no integrated and centralized organization for drainage. In some states, like Haryana, subsurface drainage is considered to be a measure for land reclamation and is executed by the Agriculture Department [260]. The surface drainage network and the canal water supply, distribution and management are, however, under the responsibility of the Irrigation Department. Improvement of on-farm (subsurface) drainage is therefore not integrated with the improvement of the main irrigation and drainage systems and improvements in management. In other states, drainage improvements within an irrigation command are entrusted to the CADAs; outside the command areas, it is the responsibility of the Irrigation Department [150]. The case study conducted in farmers' fields in India shows that, although both male and female farmers clearly see the benefits of subsurface drainage, the current support provided by the government is insufficient to introduce subsurface drainage at a large scale. Other studies confirm that major factors affecting implementation of drainage measures by farmers are their education level and socioeconomic conditions, including the size of their holdings [86]. Furthermore, drainage provisions in the development of irrigation are generally deferred or postponed due to the paucity of funds and have to be taken up subsequently as a curative measure. Although the government stresses the need to incorporate a participatory approach in the management of water resources [89], participatory approaches in subsurface drainage in India are still in a premature stage of development.

In Pakistan, drainage is generally executed within the canal irrigation commands. The drainage projects are contracted to a special project organization under the authority of WAPDA. The WAPDA is responsible for the coordination of design, construction and initial operation of the engineering works, after which the PIDs take over O&M [150]. As the drainage fees cover only around 20% of the actual expenses on O&M, the financial burden of operating and maintaining the public tubewell systems gradually became too much for the PIDs. To overcome these problems, the irrigation and drainage sector was reformed and in 1997 PIDAs were established in all four provinces [37]. System management is to be decentralized and farmers are to take part in the system development and to take over O&M through the creation of AWBs and FOs. However, the establishment of FOs and AWBs is hampered by (i) lack of farmers' involvement in policy reforms, (ii) the weak legal framework (the PIDA Acts) for implementing reforms, (iii) a lack of knowledge within the FOs and AWBs to develop and implement strategies to deal with problems and (iv) reluctance to make the shift from engineering to institutional solutions. On the other hand, there is sufficient awareness among the farmers about the benefits [122]. They also realize that the drainage systems are not adequate. In the non-saline areas, 80% of the farmers perceive the existing drainage system as inadequate; in saline areas, this percentage is slightly better at 60% [137]. Furthermore, farmers are not very eager to cooperate with government agencies [73]. In the latest Drainage Master Plan [301], the Government of Pakistan recognizes and encourages stakeholders participation in project preparation, construction and O&M, but the institutional reforms to achieve this participation have not yet been established.

It can be concluded that in all three countries the government is the driving force behind the installation of subsurface drainage. The organizational setup is purely top-down. Farmers' involvement only starts after the subsurface drainage system has been installed; only then they are asked to become involved in O&M. This top-down approach does not encourage farmers to take up their responsibilities. A more service-oriented approach has been promoted since the 1990s [134], but has not gained much of a foothold in practice.

7.2   How can the integration of irrigation and drainage be improved?

Irrigation is a strategy for enabling farmers to cope with inadequate and unreliable rainfall [239]. In irrigated agriculture, drainage is a strategy for enabling farmers to cope with irregular rainfall and to safeguard investments in irrigation by removing excess water and salts brought in by the irrigation water. Irrigation in fact brings water twice, firstly to meet crop water requirements and secondly to provide leaching of the salts brought in with the same irrigation water [236]. In irrigated agriculture, subsurface drainage is closely linked to field irrigation and surface drainage practices. Drainage is a tool for enhancing the soil capacity to act as a storage room, especially when there are options for operational control. In the irrigation season, operational control of the drainage system can be used to maintain the water table at a higher level. Thus, subsurface drainage can be considered as irrigation modernization in the sense that it saves on water use. In the off-season, the lower water table reduces the soil evaporation and thus the salinization of the root zone, making subsurface drainage an important tool for maintaining the soil quality.

The role of irrigation differs slightly between Egypt, India and Pakistan. In Egypt, where rainfall is negligible, crop production depends entirely on irrigation. In India and Pakistan, irrigation is practised to protect agriculture against the vagaries of rainfall. In these countries, the irrigation systems are not designed to cover the full crop water requirement but work on the principle that the available water is spread over a large area: 'protective' irrigation [117]. In some years not enough water is available and leaching requirements cannot be fulfilled. In all three countries, irrigation modernization will further affect the drainage water quantity and quality. In Egypt, field irrigation efficiency is quite high, but there are still possibilities for increasing irrigation efficiency through controlled drainage without jeopardizing the leaching requirements [284]. This will reduce subsurface drainage rates and total salt loads. In India and Pakistan better control of both irrigation and drainage at field level can greatly enhance water efficiency, especially if the monsoon rains are better utilized for leaching, which will affect the quality of the drainage water.

The water balance studies done in Egypt and India show that controlled drainage can also improve irrigation efficiencies. Improving irrigation efficiency means that drainage discharges and total salt load will decrease, but the quality will deteriorate as a result of the increasing concentrations of salts. The case studies conducted in India and Vietnam shows that drainage water is not only polluted with the salts brought in by irrigation, but also with dissolved nitrates and pesticides, herbicides and fertilizers and by domestic and industrial waste water. Irrigation modernization that may be initiated to make more water available for irrigation development downstream may therefore have negative externalities. Reuse of drainage water can supplement irrigation water deficiencies and can be a viable option for minimizing disposal needs. Reuse can be practised at farm, project and regional level [196]. At farm level, drainage water can be reused when it is of good quality. Farmers can pump irrigation water directly from the open drains or use shallow wells to pump groundwater. At project and regional level, drainage water can be pumped back into the irrigation system, where it is mixed with better quality irrigation water. The quantity and quality of both the irrigation and drainage water determine how much drainage water can be reused. Drainage water, however, can never be completely reused because the salts that are imported with the irrigation water have to be exported out of the area. Another complicating factor is the increasing use of waste water for irrigation. The impacts of reusing drainage and waste water on catchment hydrology, including the transport of salt loads, are still insufficiently understood [92].

An adequate and well-maintained surface drainage system is a prerequisite for the proper functioning of any subsurface drainage system. Surface drainage systems remove the excess irrigation water or rainfall as surface runoff. It is impractical and too expensive to design a subsurface drainage system to remove this excess water, especially in a monsoon type of climate. Like irrigation, the role of surface drainage also differs between Egypt, India and Pakistan. In Egypt, surface drainage is only needed to remove excess irrigation water and for farm management practices in rice fields. Consequently these systems have only limited capacity and are frequently used at their design capacity. In India and Pakistan, the main function of the surface drainage system is to remove excess rainfall during the monsoon season. Because of the high variability of the rainfall, these systems are hardly ever used at full capacity. This makes O&M extremely difficult [103]. Experiences in Pakistan show that improving surface drainage can reduce the need for (the more expensive) subsurface drainage.

The case studies conducted in the farmers' fields in Egypt and India clearly show that, besides improving irrigation efficiency, supplementary measures in soil and water management, such as gypsum application and the introduction of salt-tolerant varieties, can enhance the positive effects of subsurface drainage.

7.3   What are the main challenges in making subsurface drainage work?

The extent of the waterlogged and salt-affected areas show that, although subsurface drainage systems are technically sound, cost-effective and generally appreciated by the farmers, they are generally not promoted well enough by governments. Despite the fact that the cost-benefit ratio is favourable, it is apparently the high initial investment needs and the continuous flows of funds needed for maintenance that hamper large-scale implementation.
Without government support, farmers will not or cannot install these systems. As a result large areas are going out of production or are suffering yield losses due to waterlogging and salinity. There are several other reasons why drainage needs to receive special attention:

- in small-scale irrigation, drainage is always a joint effort. Water infrastructure in arid and semi-arid conditions is traditionally based on the water supply situation. Disposal of excess water requires a complementary infrastructure that invariably serves a multitude of users. Drainage therefore requires the cooperation of stakeholders, which makes it more difficult to organize. The case studies conducted in Egypt, India, Pakistan and Vietnam show that farmers are willing to cooperate, but that an appropriate organizational setting is often lacking;
- the boundaries of drainage units generally do not coincide with the boundaries of irrigation units. This applies to both the command area and catchment area level as well as the field level. The case study on the modified system in Egypt show that, by matching irrigation and drainage units, considerable savings, especially in water savings, can be made;

- the institutional set-up is complex and enforcement of rules and regulations is difficult. In contrast to irrigation, where direct benefits to stakeholders are involved, rules and regulations for drainage are much more difficult to enforce. Drainage fees need to be collected and, unlike the irrigation supply system, it is difficult to disconnect unwilling customers. The case study on participatory research in Vietnam shows that the existing institutional setup, often based on irrigation system layout, needs to be modified or adapted as to improve drainage efficiency;

- drainage is at the end of the pipeline. The case studies from India and Vietnam show that drainage systems not only discharge excess water and pollutants resulting from irrigation and agricultural practices, but also waste water from rural industries and rural villages. As such, drainage may pose serious threats to downstream water users. The experiences from Pakistan show that it is extremely difficult to retain excess water in the area where it is generated. Often, combinations of treatment, local reuse and accepting the export of pollutants will be the final solution that stakeholders need to agree upon;

- disposal of drainage water creates off-site externalities. Drainage water discharged back into the river from which it was originally obtained as irrigation water has a higher salt content and is also often polluted with residues of fertilizer, pesticides and waste water from villages, cities and industries. The case studies from Egypt, India and Vietnam show that, while upstream users benefit from disposal, downstream users, or society as a whole, bear the cost. This calls for state regulation;

- reuse of drainage water. Although drainage water from irrigated lands has a higher salt concentration than the irrigation water from which it originates, its quality may still be good enough for reuse in a downstream area. Drainage water can supplement freshwater resources, sometimes only after mixing. In the end, however, the salts have to be removed from the area. The examples from Egypt and Pakistan show that this disposal is easier in the downstream part of the river basin. Results obtained with basin-management simulations confirm these findings [237]. Increased irrigation efficiency and pollution of drainage water, however, put additional constraints on this reuse;

- high investment costs versus long-term benefits. Investment costs in drainage are only a fraction of the investment costs in the irrigation infrastructure (usually between 10 and 30% of the investment cost for irrigation). Nevertheless, investment costs are high and full benefits often accrue only after a few years. Salinity build-up is a slow process and subsistence farmers normally do not have the resources to invest for the benefit of the next generation.

7.4   Improving subsurface drainage practices: the way forward

7.4.1   The state of the art in subsurface drainage

Over the past 25 years, the role of subsurface drainage in arid and semi-arid regions has changed from a single-purpose measure for controlling waterlogging and/or salinity into an essential element of integrated water management under multiple land use [219]. The way subsurface drainage systems are implemented, however, has not changed much. They are often designed and implemented by government agencies or for government agencies, with the users, the small farmers, having little responsibility and making little input. This top-down approach generally involves standardized designs that take little or no account of (i) the degree of waterlogging and salinity, (ii) the preferences of the farmers, (iii) the capacity of the farmers to maintain and operate the system, (iv) supplementary measures in soil and water management needed to enhance the positive effects of subsurface drainage, such as the application of gypsum, the introduction of salt-tolerant varieties and irrigation efficiency improvements, (v) changes in land use and the corresponding changes in drainage requirements, and (vi) the quality and quantity of the drainage effluent.

The Comprehensive Assessment of Water Management in Agriculture, a critical evaluation of the developments in the water sector over the last 50 years by a broad partnership of practitioners, researchers and policymakers, calls for '*a change in the way we think about water and agriculture*' [145]. It makes a strong pledge '*to abandon the obsolete divide between irrigated and rainfed agriculture, to consider agriculture as an ecosystem and to recognize the importance of preserving the natural resource base on which agricultural productivity rests*'. Four reasons to invest in irrigation are presented: (i) to reduce poverty in rural areas, (ii) to keep up with global demand for food, (iii) to adapt to urbanization, and (iv) to respond to climate change. Although the report recognizes the role of drainage (Figure 7.2), it is surprising that, besides the remark that '*investments in drainage are likely to continue at fairly modest levels*', the role of drainage in irrigated agriculture is not addressed. In the following sections it will be argued that this is an omission: the role of subsurface drainage in irrigated agriculture will be elaborated showing that in addition to the four reasons stated above, there is a fifth reason to invest in water and agriculture: subsurface drainage to protect investments in irrigated agriculture.

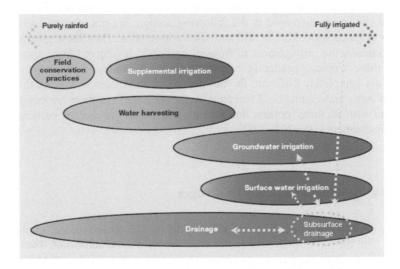

Figure 7.2   Diverse options for agricultural water management along the spectrum from purely rainfed to fully irrigated [145] (Note: the role of subsurface drainage in irrigated agriculture, the subject of this research, has been added to the figure)

A study reviewing the completed and active drainage projects financed by the World Bank identifies six main challenges facing drainage: (i) lack of an integrated approach, (ii) threats to drainage system sustainability, (ii) governance and institutional constraints; (iv) weak participatory planning and private sector role; (v) lack of focus on poverty, and (vi) policy and legal constraints [296]. The Drainage and Integrated Analytical Framework (DRAINFRAME) approach developed by the World Bank's Agriculture and Rural Development Department is a recent and promising attempt to look at agricultural drainage from an integrated natural resources management perspective [11; 296]. The DRAINFRAME approach recognizes three main settings: (i) the biophysical environment, (ii) human society, and (iii) the institutional setting. The application of the DRAINFRAME approach in Egypt and Pakistan has shown that it offers a useful approach and methodology for analysing water management situations in an integrated manner and can offer useful contributions to the project planning cycle [230]. It was also concluded, however, that the approach needs a more systematic elaboration of the stakeholders and a mature methodology for evaluating the institutional setting of water management situations.

The original DRAINFRAME report sends five messages to the broad audience of professionals in the drainage and water management sector, planners, decision makers, governments and the international community. The first message is an invitation to dare to look at all the costs and benefits. The second message calls for attention to the potential for poverty reduction, the third calls for pragmatism and vision, the fourth emphasizes the value of 'learning by doing' and the fifth is an appeal to all to promote an integrated approach. An approach to integrate the biophysical, socioeconomic and institutional aspects calls for a holistic view: 'generalization as specialization' [94]. This is a complex issue, but in line with the idea that the IWRM is a process of change: a process which starts from small beginnings '*start somewhere: do nothing is not an option*' [83]. This challenge has been taken up in the following sections. It should be mentioned here that the lessons learned and challenges discussed in the following sections are not restricted to subsurface drainage in irrigated agriculture in arid and semi-arid regions. They show great similarities with, for example, the management of tropical peatlands, a subject I have also studied in depth. In these tropical peatlands, where the socioeconomic, soil and hydrological conditions are completely different, the problems and challenges in solving them are the same: good water management is a prerequisite for wise use[14]. In these regions good water management also includes options for operation control, stakeholder participation and capacity development [199; 200; 208].

7.4.2   Institutional and policy challenges

Subsurface drainage systems, like irrigation systems, are sociotechnical entities, which means that both the physical and the social structure have to be addressed. Two levels can be distinguished: (i) the higher or governance level and, (ii) the daily O&M level. At the governance level decisions have to be taken for the longer term and on broader issues: rules and regulations on how to develop and implement subsurface drainage systems and

---

[14] Wise use is defined as use for which reasonable people, now and in the future, will not attribute blame [116].

how to organize their day-to-day O&M. Although at this level more farmers' participation is required, the role of government remains crucial since drainage requires a regional infrastructure and not only serves the farmers but also the other inhabitants in the area. These other inhabitants also benefit from the drainage system, as drainage not only improves agriculture, but also the environment and health conditions [164; 196]. In general, however, these other stakeholders do not pay any fees. The case study on participatory research shows us that because in rural areas non-agricultural land uses are on the increase, the role and responsibilities of the non-farmer community is becoming more important [204]. The present role of the government is, however, top-down and does not encourage stakeholders to take up their responsibilities. This attitude needs to be changed as subsurface drainage is a joint effort: in smallholder agriculture, a subsurface drainage system always serves more than one field/farmer. Groundwater flow does not stop at the field boundary; if only one or a few farmers install a subsurface drainage system in their fields, their neighbours have a free ride. Furthermore, subsurface drainage also requires a regional infrastructure to safely discharge the drainage water. This means that farmers have to cooperate with each other, with other stakeholders living in the area and with regional agencies.

In consultation with the stakeholders, governments would have to draw up a drainage policy that includes a time-bound action plan to reclaim the waterlogged and salt-affected lands and to safeguard other irrigated lands against these problems. An important policy measure could be a government decision to give equal importance to the reclamation of waterlogged and salt-affected areas in irrigation schemes and to the creation of fresh irrigation potential or its utilization. The current practices in India and Pakistan teach us that another policy decision is required to provide for drainage-related needs in new irrigation projects at the time of inception of the projects. In other words, the main drainage systems would have to be designed and implemented in such a way that (subsurface) drainage systems can be installed at a later date. Experiences in Egypt show that a high-level Advisory Group on Drainage that acts as a think-tank and suggests remedial measures on a case-by-case basis from time to time is an excellent tool for optimizing institutional and policy options.

A 'drainage industry', required to manufacture or supply the drainage materials and installation equipment, will only develop where there is a long-term national drainage programme for which financing is guaranteed [54]. As the benefits often go beyond the direct interest of the farmers concerned, the government would also have to finance or prefinance all or some of the costs. Funding of such programmes would have to come from the government and could be recovered in interest free instalments along with the land revenue and irrigation water charges. The cases in India show that farmers are willing to pay their share, although they have limited financial resources, especially those farmers whose land has gone out of cultivation due to waterlogging and salinity [203]. Reclamation of waterlogged and salt-affected soils, although cost-effective, seems to be beyond the reach of the small and marginal farmers as far as the initial investments are concerned.

Experiences from all over the world show that in irrigated agriculture full financial cost recovery is never achieved and is even rarely a realistic objective [250]. Governments need to be convinced of the need to invest in drainage. To convince governments and the general public of the importance of water, the concept of 'water footprints' has been developed [96]. Water footprints show the extent of water use in relation to public consumption. The water footprint of a country is defined as the volume of water needed for the production of the goods and services consumed by the inhabitants of the country [95]. The water footprint includes three components: the blue, green and grey water footprint. The last is the volume of water that is required to dilute pollutants to such an extent that the quality of the water remains above the agreed water quality standards. Although water quality is included in the calculation of water footprints [52], surprisingly enough the required leaching of salts brought in by the irrigation water is not considered [96]. Including the quality of irrigation water will emphasize the role of drainage, but at the same time increase the size of the water footprint. For irrigated agriculture in Egypt, India and Pakistan, where the quality of the irrigation water diverted from the Nile, Indus and Ganges is good (EC_i around 0.3 dSm^{-1}), this increase will be between 4 and 15%, depending on the type of crop (Table 7.2). Conjunctive use of surface and groundwater, which results in a poorer water quality (e.g. an EC_i between 1 and 2 dS m^{-1}), will increase the water footprints even further. This increase in the value of a footprint is the price one has to pay to sustain irrigation and drainage is the tool to achieve this.

Table 7.2   Additional irrigation water requirements to leach the salts imported by irrigation water: example from Gujarat, India [202]

Crop	Crop water requirements (mm)	Additional irrigation water required to leach the salts[a] (mm)		
		Canal water $EC = 0.3$ dSm^{-1}	Groundwater $EC = 1.0$ dSm^{-1}	Groundwater $EC = 2.0$ dSm^{-1}
Banana	1,920	288 (15%)	960 (50%)	1,920 (100%)
Cabbage	480	40 (8%)	133 (28%)	267 (50%)
Cotton	280	6 (2%)	20 (7%)	40 (14%)
Rice	1,200	51 (4%)	171 (14%)	343 (29%)
Sugarcane	1,500	132 (9%)	441 (29%)	882 (59%)
Sorghum	240	9 (4%)	30 (13%)	60 (25%)

[a]   based on maximum soil salinity level without yield reduction [74]

7.4.3   Increased stakeholder participation

Stakeholder participation in subsurface drainage is essential. Among the reasons for this are that (i) subsurface drainage involves more than one farmer, (ii) waterlogging and salinity problems are location-specific, but (iii) drainage also requires a regional infrastructure to safely discharge the excess drainage water, (iv) government involvement in planning and implementation is required, but in the end it is the farmers who have to operate and maintain the system, and (vi) to increase cost recovery the financial participation of all stakeholders must be raised to the highest acceptable price.

The way land consolidation was organized in the Netherlands is an example of how stakeholders can be involved. In the Netherlands land consolidation was an instrument used to modernize Dutch agriculture after the Second World War [263]. An essential element of the land consolidation planning was that a majority of the stakeholders had to agree on the plans before implementation could start. This was a slow and lengthy process of negotiations that often took several years. It became even more complex when other functions, such as nature conservation and recreation, became more important. To deal with the increasing complexity, the Land Consolidation Act was replaced by the Land Use Act in 1985. Successively new methods for participation and social learning, like community of practice, multi-stakeholder platforms, etc., were introduced.

The subsurface drainage practices in Egypt, India and Pakistan show that the government wants the farmers to take responsibility for O&M. To achieve this, stakeholder participation by farmers and also the other stakeholders, will have to increase at various phases in the implementation process (Figure 7.3).

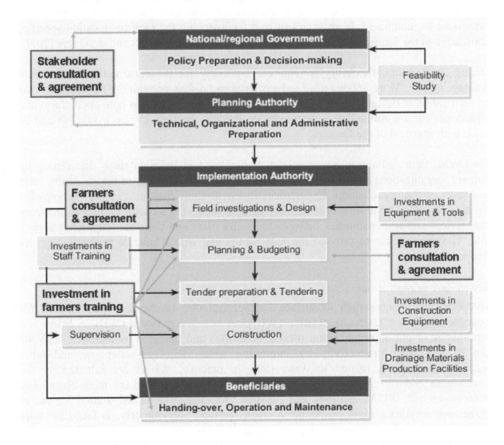

Figure 7.3   Participation by farmers and other stakeholders has to be embedded in the implementation process

During preparation and planning, more attention has to be paid to the identification of the stakeholders and their needs and preferences. Lessons learned in Pakistan show that planning is a dynamic process [301]. The Kolleru-Upputeru case shows that participatory modelling is a useful tool for identifying the needs and preferences of the different stakeholders and creating consensus on an integrated development approach. The case on stakeholder participation in drainage research shows that a Participatory Learning and Action (PLA) approach, including the stakeholders, is an effective and efficient method for assessing the need for drainage and obtaining agreement between the stakeholders on follow-up actions.

During field investigations and design, better interaction is needed between the stakeholders and the various government agencies involved in rural development. Snellen, in an analysis of the Indian irrigation sector, concludes that one of the main problems in this sector is that '*Irrigation professionals lack the attitude, knowledge, and skills to provide water services to agriculture and to protect the land and water resources*' [239]. This applies even more to drainage. The case study on capacity development shows that an integrated approach, in which research, training and education and advisory services (or extension) are combined, is an excellent tool for matching the tacit (or location-specific) knowledge of the farmers with the explicit knowledge of the planners and designers [207].

During construction, flexibility is required, especially for the construction of the field drainage system. More options need to be offered and discussed with the farmers: the case on verification of design criteria in farmers' fields in India shows that both open drains and pipe drains (or a combination of both) can do the job; the final choice would have to be made with approval of the farmers.

For O&M, farmers have to be organized. Experiences in Pakistan show that setting up farmers' organizations is not easy: the stakeholders have to agree on the structure, rules and tasks of the organization. Experiences in Egypt show that stakeholder participation needs to be scaled up to district level. Service agreements are a useful tool for formalizing the services and responsibilities between the water users and the water supplying agency [58]. Government has to provide the legal status and would have to invest in capacity development.

Stakeholder organizations

To operate and maintain their subsurface drainage systems, stakeholders can cooperate, for example in a Water Users Association (WUA) or in a Water Board (WB). A WUA is a semi-public or public organization owned, controlled and operated by members with the aim of increasing productivity by improving water delivery, water use and other organizational efforts related to water [99]. In practice, WUAs are initiated by the government and the initiative does not originate from the farmers themselves. The experiences with WUAs in Egypt and Pakistan show that establishing a good legislative framework remains a main challenge; a challenge that, unfortunately, is faced by more countries, not only in the developing world [97] but also in Europe [128]. In the Netherlands, Water Boards have the legal power to levy taxes from the inhabitants for drainage services (including water treatment) [58; 160; 261]. These Water Boards are decentralized functional government authorities and can focus fully on water governance, which is therefore to a large extent immune to political whims [255]. Water Boards, however, need extensive legislation.

Farmers are only willing to participate if they gain in economic or non-economic terms by doing so. Group formation can be supported by financial and technical assistance and will be more successful when the cooperation is based on existing social organizations, for example Water Resources/Irrigation Departments in association with irrigation water users associations or groups or village *panchayats* (in India) [86]. This is in line with the lessons learned in the Netherlands, where in the 1970s, the reform of the Water Boards showed that it is better to reform existing institutions than to abolish them because they are already deeply embedded in society [238]. This is in line with experiences from elsewhere, e.g., a review of WUAs in Peru showed that it is essential that the diverse local realities and the various managerial customs, methods and styles are taken into consideration [97].

In Egypt, an analysis of experiences with WUAs in irrigation showed that they are likely to have a greater impact on water control when water supply is unstable and at the same time a process of crop diversification and intensification is underway [99]. This is an encouraging experience as elsewhere these two changes – a more unreliable water supply (due to increasing demand for water in combination with climate change) and more intensive land use – are also likely to take place. The findings in Egypt are supported by the lessons learned in India, where farmers in the Uppugunduru pilot area, who already cooperated on irrigation, relatively easily took over responsibility for the O&M of the subsurface drainage system.

The case study on drainage pilot areas in India shows that women, although they only play a minor role in the O&M of subsurface drainage systems, clearly see and feel the benefits because they reduce their work load. A flexible, bottom-up and participatory approach will help to recognize women as actors, and to identify gender concerns [302].

The case study conducted in Vietnam shows that a participatory research approach is a good strategy for analysing the existing institutional setup and to obtain agreement on improvement options. Improvement options need to be based on the following principles [204]:
- clear and transparent responsibilities;
- responsibility at the lowest possible level;
- include stakeholders;
- charge stakeholders;
- need for monitoring;
- need for capacity building.

Community of practice

The 'community of practice' social learning technique is an accepted part of organizational development. It is a type of action research in which stakeholders with a common interest work together to share ideas, find solutions and build innovations [262]. The aim is to make the tacit knowledge of the stakeholders explicit and to combine it with explicit knowledge from elsewhere (for example research knowledge) to create new knowledge. The new knowledge, which now incorporates the location-specific characteristics (both physical as well as institutional), is used to plan drainage interventions. Community of practice is an effective tool to promote the flow of knowledge between researchers, policymakers and resource managers [210]. Both the case study on participatory drainage research and the case study on participatory modelling have elements of community of practice.

Multi-stakeholder platforms

In practice, IWRM is rather difficult to implement as authorities are in general reluctant to accept alternative systems of governance [288]. Multi-stakeholder platforms (MSPs) are a new form to overcome this problem. A MSP is a decision-making body comprising of different stakeholders who perceive the same resource management problem, realize their interdependence for solving it, and come together to agree on action strategies solving the problem [285]. Experiences in Europe, South Africa and Latin America show that MSPs can made a difference, as long as there is (i) an appreciation that empowerment and transformation of stakeholders are not automatic, and (ii) a recognition that many people are competitive and driven by self-interest rather than predisposition to cooperate and collaborate. Verhallen et al [276] show that MSPs can enhance accountability and adaptivity to autonomous social and environmental challenges, as well as the tensions generated by interventions to counter them [276]. These authors warn, however, that the multiple pitfalls are '*smokescreen participation*' and '*talking shop*'. They stress that MSPs should be organised in co-production with the stakeholders themselves: '*do it well, or don't do it at all*'.

Participatory learning and action

Another instrument in stakeholder participation is participatory learning and action (PLA). PLA is an approach for joint learning and planning with communities [87; 252]. It entails a set of participatory tools and visual methods such as mapping, making time lines, transect walks, constructing problem trees, ranking activities and making Venn diagrams [264]. PLA goes beyond mere consultation and promotes the active participation of communities in the issues and interventions that shape their lives. It enables local people to share their perceptions and identity, and prioritize and appraise issues from their knowledge of local conditions. By combining the sharing of insights with analysis, PLA provides a catalyst for the community to act on what is uncovered. The case studies conducted in India and Vietnam show that PLA can be a useful tool in drainage planning.

Participatory modelling

Implementation of national water policies at local level is a major struggle. Organizational complexity and involving stakeholders are the most important constraints and at the same time the most important conditions for success [84]. Models are useful to get a better understanding of complex water management problems with many stakeholders and limited data records [197; 204]. Furthermore, the cases in India and Vietnam show that participatory modelling is a useful tool for finding a balance between top-down control and bottom-up collaborative planning. The planners benefit because they can use the location-specific knowledge of the stakeholders to develop their models. Both the planners and stakeholders get a better understanding of the location-specific problems (both physical and institutional) and their interrelated complexity. This participatory modelling approach has also proved to be a useful method for creating mutual understanding and developing the outlines of an integrated approach.

Progressive farmers

In India, the concept of 'progressive farmers' is used to increase participation [211]. A number of farmers are selected or step forward themselves to be trained in new or innovative subsurface drainage techniques and they are supported during implementation.

The examples from Gujarat and Andhra Pradesh, where farmers or groups of farmers took the initiative to solve their waterlogging and salinity problems, show that this approach can be successful in introducing subsurface drainage practices in a region.

7.4.4   Drainage system requirements

Restoration of the natural drainage network

In irrigated areas, the (natural) drainage capacity is often disrupted by the construction of the irrigation canal networks, roads or railways. Restoration of the natural drainage capacity will considerably reduce the need (or intensity) of the subsurface drainage and will furthermore be needed to provide safe outlet conditions. The availability of a good outlet for safe disposal of the drainage effluent (including the leached salts) is a prerequisite for making subsurface drainage a success. In landlocked areas, 10–17% of the land area is needed to store water in evaporation /storage tank and to dispose of it at a later stage [196].

Recognizing the multiple use of drainage systems

Drainage systems are generally designed and managed as if they only serve to dispose excess water and salts. The case studies show that these systems are also used to dispose of sewage from rural settlements and waste water from rural industries. As in irrigation systems, recognizing the multiple uses of the drainage systems is critical for better design and management [140]. Although dealing with other stakeholders will make the institutional arrangements more complex, it also offers opportunities as the value of the system has been undervalued. Recognizing the rights and obligations of all users will lead to a more equitable and socially sustainable system.

The need for a location-specific approach

Drainage measures need to be adapted to the prevailing soil and hydrological conditions [231], which generally vary from place to place [39]. The experiences from India and Pakistan show that drainage measures also need to be tuned to the prevailing socioeconomic conditions of the farmers [203]. As not all fields have the same need for subsurface drainage, more attention would have to be paid to the degree and extent of the waterlogging problems and the preferences and capacities of the stakeholders to pay for and maintain the drainage interventions. For this, local knowledge is indispensable. This calls for a more flexible, participatory approach.

The need for operational control

A well established paradigm for agricultural drainage is controlled drainage, or draining only when needed [282]. The currently installed systems in arid and semi-arid regions, however, have limited options for operational control. The water balance studies conducted in the pilot areas in Egypt and India show that there is still significant potential for improving field water management practices by controlled drainage in combination with improved irrigation water efficiency.

In Egypt the systems are based on average cropping patterns and conditions, and so most of the time excessive drainage occurs. As a result each year approximately 18.5 billion cubic metres of water are drained from areas provided with subsurface drainage systems [7]. The case study on the modified layout for areas with rice in the cropping pattern shows that better operational control can reduce both irrigation water requirements and drain discharges [66]. Controlled drainage can also reduce irrigation water supply of the non-rice crops like wheat and maize without sacrificing yields [1]. Model simulations show that application of controlled drainage has the potential to maintain and even increase yields while increasing irrigation water use efficiency by 15–20% [284].

In India and Pakistan, which have a distinctive monsoon season, controlled drainage is a logical option for saving irrigation water during the dry season. Field research conducted in Andhra Pradesh, India, showed that pumps have to be operated for only about 10–15% of the time to maintain a favourable salt balance [103]. In drained areas of Pakistan where a deep water table is maintained, farmers sometimes complain about the increased need for irrigation water [177]. A shallow water table, especially in the fine soils of the Indus plains, is capable of water delivery to the crops through capillary rise. In areas with an 'acceptable' groundwater quality there is no need to maintain a deep water table. Controlled drainage can reduce the nitrate concentration in the drainage effluent [10]. A study conducted in Andhra Pradesh, India, shows that subsurface drainage causes approximately 3–20% loss of the total applied nitrogen [226]. A negative consequence, however, is that the lower drainage intensity may increase groundwater pollution [226].

Climate change

More control may also be required in response to climate change. As temperatures increase irrigation water requirements will increase. Moreover, precipitation patterns are shifting, and rainfall events may become more extreme [268]. More storage is required not only for irrigation, but also for drainage. As storage mainly has to be found in or on the soil, adaptation has to take place at local level. Coping with natural climate variability is nothing new for water management and farmers are already used to adaptation and risk assessment, even in irrigated agriculture [239]. Their 'tacit' knowledge can be used to respond to climate change. What cannot be predicted with certainty is precisely where and when weather extremes will occur, or just how extreme they will be. This creates uncertainty for both the farmers and the authorities and calls for stakeholder participation and a flexible approach.

The need for a more flexible approach

The current installation practices are rather rigid and top-down, but changing land use requires a more flexible approach. The case studies conducted in India and Vietnam shows that crop intensification and diversification and non-agricultural land use require a more intensive drainage system. In general the implementation of subsurface drainage systems follows a three-step approach: (i) construction or remodelling of the main drainage system, (ii) construction or remodelling of the pumping stations, and (iii) construction of the field drainage system. Flexibility and farmers' participation are generally less necessary in the first two steps than in the last step. Care need to be taken that main systems and pumping stations do not restrict flexibility with the construction of field systems. Flexibility implies that monitoring and evaluation becomes even more important. Up-to-date information is needed to update the design and implementation practices, not only to evaluate and improve the existing subsurface drainage practices, but also to act on climate change and/or land use changes.

Boundaries of the drainage unit

Research conducted in Egypt clearly shows that the boundary of a drainage unit preferably would have to coincide with the boundaries of irrigation units [7]. This is confirmed by research findings in Andhra Pradesh, India, where participatory drainage management was especially successful in the Uppugunduru pilot area. In this area farmers have been able to manage the subsurface drainage system successfully since July 2003, when they took over the system at the end of the IDNP, mainly because they already cooperated among each other on irrigation [213].

7.4.5   Capacity development

Capacity development, in the context of IWRM, represents 'the sum of effects to nurture, enhance and utilize the skills and capabilities of people and institutions at all levels, so that they can work together towards the broader goal of IWRM' [83]. Capacity development on drainage and water management for salinity control in irrigated agriculture involves engineers, agronomists, soil scientists, socioeconomists, extension scientists, extension workers and farmers. Capacity building is more difficult at the management level than at the design and construction level [134]. At both levels, actions must be socially as well as technically informed, and this may require a new kind of water professional [143]. As the needs and approaches of capacity development are likely to continue to change in future [80], capacity development is as much a process as a product in which (applied) research, training and education, and advisory services (or extension) are essential elements [207]. Experiences from the Netherlands show that a sound knowledge base, for example cooperation between research and industry to improve techniques and materials, is a prerequisite for success [54]. To initiate such cooperation, government support is required because the industry and the technology will only become mature and self-sustaining after about 20–30 years.

In Egypt, India and Pakistan, research on drainage has been formalized and there is a strong link with implementation. The link with education, however, is weak; only in India do agricultural universities have a strong role in extension and dissemination. Although the need for participation by farmers' groups and relevant civil society organizations in capacity development is generally recognized [181; 299], detailed arrangements suitable for subsurface drainage practices have not yet been established. Direct interaction between researchers and farmers is an essential feature for applied research in farmers' fields [82]. Whereas in the past, technology transfer was mainly top-down, nowadays it is a more iterative process based on the principles of learning by doing (Figure 7.4). Careful diagnostic research is needed to understand farmers' systems and determine (i) where further research might efficiently and effectively help to solve problems, and (ii) where further research may have few benefits [81].

To increase the momentum of socio-economic transformations in irrigated agriculture priority should be given to empowerment of water users through training and extension

[32]. The analysis of the examples presented from Egypt, Pakistan, India and South-East Asia shows how research, training and advisory services can be linked to the knowledge creating process as described by Nonaka-Takeuchi (Figure 6.4). By following the four steps of Nonaka's knowledge creating process, explicit knowledge is internalized, for example through education and training, and then linked to tacit knowledge by socialization, for example through applied research. Bringing tacit and explicit knowledge together yields new knowledge through externalization, which in turn can be used, in combination with explicit knowledge from elsewhere and used in recommendations and by advisory services. It should be realized that capacity development is a dynamic process in which the phases of Nonaka's creating process are repeated a number of times. The cases show that capacity development is much enhanced when there is a long-term partnership [207]. Empirical evidence from partnerships in Mali and Egypt also strongly suggests positive impacts on capacity development: research capacity was built and enhanced [30].

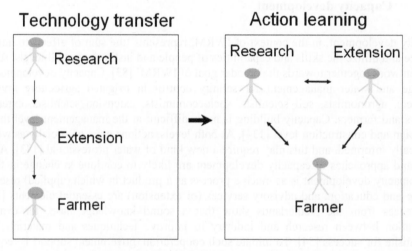

Figure 7.4   The role of research has changed from top-down to a more iterative process based on the principles of 'learning by doing' [209].

Applied research

Applied research in farmers' fields is needed to adapt subsurface drainage practices to local conditions. Applied research delivers value for money. It has helped to modernize subsurface drainage practices and considerable savings have been achieved through the introduction of (i) new methods to investigate and identify areas in need of drainage, (ii) new design and planning methods, (iii) new materials for pipe drains and envelopes, (iv) improved drainage machinery and equipment, and (v) improved installation, O&M methods and practices. Finally, research is needed to improve the organization of subsurface drainage operations and institutions. This is a complex and lengthy process that needs determination and continuity. Egypt has proved that taking up this challenge can be rewarding [7]. That Egypt has nowadays one of the largest and most modern subsurface drainage programmes in the world can, to a large extent, be attributed to these capacity development activities. This is especially remarkable since the developments in Egypt took place in a relatively short period of 40 years, during which manual installation practices were almost completely mechanized, including the introduction of new materials [149]. An effective research management system includes the capacity and authority to set research priorities and enforce them through the allocation of resources and the critical evaluation of research activities [82]. This was the case in Egypt, where the Advisory Panel supported policymaking and strategic planning at ministerial level and the research by the DRI was conducted under the authority of EDPAP, the implementing agency and the guidance of the panel.

The right blend of research methods

The case study on the modified system for rice areas in Egypt illustrates the various levels at which research can be conducted: (i) research in farmers' fields, (ii) research in experimental plots/stations, (iii) extension and (iv) international/regional networking. Because the concept of the modified system is based on theories and practices on controlled drainage developed outside Egypt, it had to be verified under the prevailing conditions in the Nile Delta before it could be introduced. Research was conducted at four levels: (i) a desk study to adapt the concept to Egyptian conditions, (ii) fully controlled experiments at three experimental stations, (iii) in-depth studies in farmers' controlled fields, and (iv) large-scale monitoring programmes in three project areas. A foreign consultant was hired to assist with the desk study [49]. In the experimental plots, water and salt balance studies under fully controlled conditions were conducted to assess the potential of the modified system: irrigation water savings, drainage discharge rates and salinities, types and performance of closing devices, changes in soil salinity, crop yields, etc [66]. Fully controlled conditions were needed because crop yield depends on so many factors that without control over the inputs conclusive results are very hard to obtain. In the pilot areas, the results obtained in the experimental fields were verified under farmers' controlled conditions. The farmers had full control of the inputs and outputs, such as irrigation water application, opening of the blocked subcollectors, etc. O&M practices were agreed upon between the farmers, the Drainage Authority and the Ministry of Agriculture. Finally, large-scale monitoring programmes were conducted to verify the crop consolidation practices and the operation of the closing devices [63]. The same research activities were also conducted in areas drained according to the traditional system. By doing so, the influence of autonomous developments, like new crop varieties and farming practices, could be accounted for.

A stepwise approach to training

A stepwise approach in capacity development for engineers and scientists, as used in India, is recommended [102; 207]:

- basic training by attending regular training courses at specialized centres;
- tailor-made courses on more advanced subject matters;
- project expert meetings to exchange experiences and to synchronize research activities;
- collaborative research programmes at universities and/or research organizations to provide in-depth training and capacity development on specialized subjects;
- in-service training to introduce new research methods and tools;
- study tours to be informed on practices from elsewhere.

Learning by doing

The never-ending learning loop of translating tacit into explicit knowledge requires a continuous exchange of knowledge (Figure 5.4). This exchange of knowledge can be enhanced by using tools like in-service training, train-the-trainers programmes and vocational training centres. In in-service training and train-the-trainers programmes experienced instructors and trainees visit and train field staff. At vocational training centres, such as DTC at Tanta, the instructors are engineers with many years of experiences in drainage practices [149; 201].

Dissemination and uptake

Subsurface drainage practices, even when based on location-specific conditions, always have to be fine-tuned in the field. A study conducted by the International Programme for Technology and Research in Irrigation and Drainage (IPTRID) and the World Bank on modern water control and management practices in 16 irrigation projects in ten developing countries concluded that most design and operation solutions to improve water management are simple, but the institutional problems are not [45]. The main constraints was the lack of location-specific knowledge needed to resolve specific issues ('the devil is in the detail'). The reason was simple: the farmers had no control over the topics being discussed, but at the same time they were supposed to do the final fine-tuning in their fields. Training of farmers and field staff on innovative water management practices is a prerequisite for success. This type of training can only be successful if the specific training needs of these farmers and staff are taken into consideration. As such, training has to be based on knowledge of the social environment [274]; training manuals for extension workers should not focus on theories but on practical implementation. An example of how explicit knowledge can be transformed to link it with tacit knowledge is FAO Training Manual no. 9, *Drainage of Irrigated Lands* [198]. This manual provides guidelines to help field assistants in agricultural extension services and irrigation technicians at the village and district level to increase their ability to deal with farm-level water management issues. The manual is based on the theoretical background of field drainage as presented in ILRI Publication 16 *Drainage Principles and Applications* [191], but because the target group does not have to design or implement subsurface drainage systems, the design principles, which in Publication 16 are presented in equations and symbols, have been replaced by photos and figures illustrating practical problems and solutions. In Egypt, limited documentation materials and the fact that most documentation was in English made it difficult for the engineers and farmers to comprehend and adopt research findings [212]. The research done in India shows that dissemination is enhanced when location-specific recommendations are translated into the local language [214; 240].

8   The way forward: enhancing the role of subsurface drainage

The overall objective of this study was to highlight the potential of subsurface drainage for improving irrigated agriculture in arid and semi-arid regions by:

- reviewing subsurface drainage practices in irrigated agriculture in Egypt, India and Pakistan;
- assessing the performance of these subsurface drainage systems, especially in relation to IWRM;
- identifying improvement options in the planning, design, installation, O&M practices in order to increase efficiency, equity and environmental sustainability;
- showing when these improvements are useful in contributing to the increasing demand for food, in safeguarding investments in irrigated and non-irrigated agriculture, and in conserving land resources.

This study may provide professionals with tools to improve subsurface drainage practices in arid and semi-arid regions by giving answers to the following questions:

- under which conditions is subsurface drainage a technically feasible, cost-effective and socially acceptable option for sustaining agriculture in irrigated lands?
- how can the integration between irrigation and drainage be improved?
- what are the main challenges in making subsurface drainage work at a larger scale?

These questions have been addressed by comparing subsurface drainage practices in Egypt, India and Pakistan and, when appropriate, supplemented by experiences from other countries. The answers to these questions can also help planners and decision makers to address issues related to an enabling environment and the changing roles of institutions in improving farmers' participation and organization.

Lessons learned

To feed their growing populations, countries like Egypt, India and Pakistan have invested heavily in irrigation. In these countries, the majority of the population is still employed in the agricultural sector and the majority of these farmers are smallholders, owning less than 1 hectare of cultivated land. To protect the irrigated lands against waterlogging and salinity, subsurface drainage systems have been or are being installed. In all three countries, the government is the driving force behind the implementation of subsurface drainage. Over the last 50 years, subsurface drainage practices have evolved from manual to large-scale mechanized installation. These subsurface drainage systems effectively prevent waterlogging and root zone salinity and consequently increase crop yield and rural income. The systems are also very cost-effective. The recent rise in the price of the major food commodity prices will increase the economic returns even further.

However, although the installed systems are technically sound and cost-effective, drainage development lags behind irrigation development and consequently a substantial part of the irrigated areas suffer from waterlogging and salinity. This is mainly because the subsurface drainage systems are designed and implemented by government, with the users, the small farmers, having little responsibility and making little input. These farmers are poor and do not have the means to invest in subsurface drainage themselves. In the adopted top-down approach the location-specific conditions and farmers preferences are hardly taken into consideration. Furthermore, the emphasis has been on the technical aspects (the physical infrastructure), while the organizational aspects (institutional infrastructure) have been largely neglected. A shift to a more service-oriented approach, as promoted since the 1990s, has not gained much of a foothold in practice.

There is, however, scope for improvement, as the study shows that:
- although farmers generally do not have the means or knowledge to implement drainage on their own, they clearly see the benefits of drainage and are willing to take up their responsibility and participate in the planning, construction and O&M, both in cash and in kind;
- involvement of the stakeholders not only enhances their empowerment, but also improves subsurface drainage practices by integrating their tacit (location-specific) knowledge with the explicit knowledge of the planners, designers and implementers;
- an integrated capacity development approach, in which research, training and education and advisory services (or extension) are combined, is an excellent tool for integrating this tacit and explicit knowledge;
- participatory learning and action approach is an effective and efficient method for involving stakeholders in the assessment of the need for drainage, creating a mutual understanding of the problems and developing an integrated approach to development;
- participatory modelling is a useful tool for creating a better understanding among the stakeholders of the complexity of the problems and effectiveness of solutions;
- there are several options for optimizing subsurface drainage systems as they are still over-designed, standardized and not attuned to location-specific conditions, neither physically nor socioeconomically;
- better operational control will enhance the integration between irrigation and drainage and will thus save irrigation water and reduce drain discharges without sacrificing crop yields.

Challenges in making subsurface drainage work

Based on these findings, I have identified the following challenges for enhancing the role of subsurface drainage:

(i)   *Balancing top-down against bottom-up.* Instead of a top-down approach, a more participatory approach is recommended, focusing on societal choice and decentralized management. Participation is needed throughout the complete implementation cycle: from planning, design and installation to O&M. This requires policy and institutional changes in which governments need to take the lead. Reform of existing institutions is better than abolishing them because they are rooted in society. It is recommended that governments, in close consultation with the stakeholders, develop drainage policies that emphasize the need to treat the reclamation of waterlogged and salt-affected areas in irrigation projects and the creation of fresh irrigation potential or its utilization with equal importance. Such a policy would have to include a time-bound action plan to safeguard irrigated lands against these problems. As the benefits often go beyond the direct interest of the farmers concerned, governments need to finance or prefinance some of the costs. The other stakeholders, including the farmers, also need to contribute, either in cash or in kind (labour). Organizing participation is a long and troublesome process: recent approaches, such as participatory learning and action, participatory research, including participatory modelling, and communities of practice, can enhance this process. This calls for a joint effort by government and the WUAs to design and implement a strategy for capacity development and institutional strengthening. This strategy needs to enhance the link between technical aspects (requiring physical solutions) and organizational aspects (requiring institutional changes).

(ii)   *From standardization to flexibility.* Instead of standardized design and implementation practices a much more flexible approach based on location-specific conditions and farmers' preferences is recommended. Integration between the irrigation and drainage network needs to be improved. The challenge is to find a balance between the individual need for drainage, which varies from field to field, and the fact that drainage at farm level is a collective activity. There is also the need for better operational control. Controlled drainage will allow farmers to optimize their on-farm water management, based on the specific conditions and their own preferences. Furthermore it enables the farmers to respond on changes in land use and/or the effects of climate change.

(iii)   *Focus on capacity development.* Capacity development is a prerequisite for success. In agricultural land drainage, capacity development is as much a process as a product in which research, training and education, and advisory services (or extension) are essential elements. In this knowledge-creating process, the explicit knowledge of the researchers can be internalized (learning) through education and training and then linked to the tacit knowledge of the stakeholders through socialization (sharing experiences) by conducting applied and participatory research. Bringing tacit and explicit knowledge together will yield new knowledge (externalization). In turn, this can again be combined with explicit knowledge from elsewhere (synthesis) to be used, for example through guidelines and advisory services, in design, implementation and operation and management. Experience shows that this capacity development process does not necessarily have to be done in one project. It is essential, however, that the three basic elements – research, education and advisory services – are applied in an integrated manner.

To be able to face these challenges, all involved parties, including the government, have to work together. Working together will lead to mutual trust between the cooperating partners, which is much enhanced when there is long-term partnership. Establishing such partnerships has been one of the Global Water Partnerships' major goals in working towards integrated water resources management. Although it should be realized that many stakeholders are driven by self-interest rather than predisposition to cooperate. It is important that financiers realize that stakeholder participation is a complex and lengthy process that can take many years and that the normally adopted project cycle of donors of 1 to 4 years is generally too short to optimize these processes. To make subsurface drainage an effective tool for solving major land and water problems, governments and financiers, must realize this and be willing to invest in such long-term processes.

I am convinced that the need for subsurface drainage will increase significantly help to meet the increasing demand for food, to safeguard investments in irrigated and non-irrigated agriculture and to conserve land and water resources. I am also convinced that the above challenges will facilitate the further introduction of subsurface drainage in irrigated agriculture in arid and semi-arid regions and through this contribute to a better, more sustainable use of the precious land and water resources in these regions. However, further research and development is still needed to meet the specific needs of emerging and least-developed countries, which each have their own specific climatic, physical and socioeconomic conditions. Furthermore, the specific needs of subsurface drainage are also changing, particularly with regard to climate change, land use changes and the quantity and quality of drainage water. These changes will require modifications in subsurface drainage practices from planning, design and implementation to O&M. It is the farmer who has to adapt his farming system to these changing needs and field scale and for the irrigation and drainage agencies for the main system. The challenge for the research and education community is to support farmers in managing their fields in a more sustainable way and to enable them to cope with these changes. Only if these challenges are met will investments in irrigated agriculture be protected, increasing its sustainability and its chances of feeding the growing world population.

References

[1] Abbott, CL, Abdel-Gawad, S, Wahba, MS, and Lo Cascio, A. 2001. Field testing of controlled drainage, and verification of the WaSim simulation model. Report no. OD/TN 102, HR Wallingford.

[2] Abdalla, MA, Abdel-Dayem, S, and Ritzema, HP. 1990. Subsurface drainage rates and salt leaching for typical field crops in Egypt. Symposium on Land Drainage for Salinity Control in Arid and Semi-Arid Regions, Cairo, 383-392.

[3] Abdel-Dayem, MS. 1986. A comparative study on the use of PVC and PE in manufacturing corrugated drainage tubing. Technical Report No. 53, Drainage Research Institute, Cairo.

[4] Abdel-Dayem, MS. 1987. Development of land drainage in Egypt. Proceedings, Symposium 25th International Course on Land Drainage Wageningen, 195-204.

[5] Abdel-Dayem, MS. 1997. DRAINMOD-S as an integrated irrigation and drainage management tool. In: WB Snellen (Ed). Towards integration of irrigation and drainage management. Proceedings Jubilee Symposium, Special Report, Wageningen, 49-63.

[6] Abdel-Dayem, MS. 2007. Drainage experiences in arid and semi-arid regions.

[7] Abdel-Dayem, MS, Abdel-Gawad, S, and Fahmy, H. 2007. Drainage in Egypt: a story of determination, continuity and success. *Irrigation and Drainage*, **56**, S101-S111. DOI: 10.1002/ird.335.

[8] Abdel-Dayem, MS, and Ritzema, HP. 1990. Verification of drainage design criteria in the Nile Delta, Egypt. *Irrigation and Drainage Systems* **4**,117-131.

[9] Abdel-Dayem, S. 2006. An integrated approach to land drainage. *Irrigation and Drainage*, **55** (299-309).

[10] Abdel-Dayem, S, and Abdel Ghani, M. 1992. Concentration of agricultural chemicals in drainage water. Drainage and Water Table Control. Proceedings 6th Int. Drainage Symposium, Nashville, Tennessee, 353-360.

[11] Abdel-Dayem, S, Hoevenaars, J, Mollinga, PP, Scheumann, W, Slootweg, R, and van Steenbergen, F. 2004. Reclaiming drainage: towards an integrated approach. Agriculture and rural development report No. 1, The International Bank for Reconstruction and Development, Washington DC.

[12] Abdel-Dayem, S, and Ritzema, HP. 1987. Covered drainage systems in areas with rice in the crop rotation. International Winter Meeting of the American Society of Agricultural Engineers, Chicago, 7.

[13] Abdel-Dayem, S, and Ritzema, HP. 1987. Subsurface drainage rates and salt leaching in irrigated fields. 3rd International Workshop on Land Drainage, Columbus, Ohio, F:1-8.

[14] Abdel-Dayem, S, and Ritzema, HP. 1990. Verification of the drainage design criteria in Egypt. Symposium on Land Drainage for Salinity Control in Arid and Semi-Arid Regions, Cairo, 300-312.

[15] Abdel-Dayem, S, Ritzema, HP, El-Atfy, HE, and Amer, MH. 1989. Pilot areas and drainage technology. In: MH Amer and NA de Ridder (Eds). Land Drainage in Egypt, Drainage Research Institute, Cairo, 103-161.

[16] Abdel Gawad, ST, Khalek, A, M.A., Boels, D, El-Quosy, DE, Roest, K, C.W.J., Rijtema, PE, and Smit, MFR. 1991. Analysis of water management in the Eastern

Nile Delta. Re-Use of Drainage Water Project Report no. 30, Winand Staring Centre, Wageningen.

[17] Abdel Ghany, MB, Lashin, I, Vlotman, WF, and El-Salahy, A. 1997. Farmers participation in the operation of modified drainage system. In: WB Snellen (Ed). Proceeding of the jubilee symposium at the occasion of the fortieth anniversary of ILRI and thirty-fifth anniversary of the ICLD (25-26 November 1996), Wageningen, 81-93.

[18] Abid Bodla, M, Hafeez, A, Tariq, M, Chohan, MR, and Aslam, M. 1998. Seepage, equity, and economic evaluations for irrigation canal lining in Fordwah Eastern Sadiqia South project. IWASRI Publication 204, International Waterlogging and Salinity Research Institute, Lahore.

[19] Advisory Panel Project on Water Management. 2003. Precious water – a celebration of 27 years of Egyptian-Dutch cooperation. Advisory Panel Project on Water Management, APP Central Office, Cairo.

[20] Ali, AM, van Leeuwen, HM, and Koopmans, RK. 2001. Benefits of draining agricultural lands in Egypt: results of five years' monitoring of drainage effects and impacts. *Water Resources Development* **17** (4), 633-646.

[21] Alterra-ILRI. 2001. Final report Netherlands research assistance project, 1988-2000. Netherlands Research Assistance Project. , Alterra-ILRI, Wageningen.

[22] Amer, MH, Abdel-Dayem, MS, Osman, MA, and Makhlouf, MA. 1989. Recent developments of land drainage in Egypt. In: MH Amer and NA De Ridder (Eds). Land Drainage in Egypt, Drainage Research Institute, Cairo, 67-93.

[23] Amer, MH, and Abu-Zeid, M. 1989. History of land drainage in Egypt. In: MH Amer and NA De Ridder (Eds). Land Drainage in Egypt. Drainage Research Institute, Cairo, 43-66.

[24] Amer, MH, and De Ridder, NA. 1989. Land drainage in Egypt. Drainage Research Institute, Cairo.

[25] Andhra Pradesh Water Management Project. 2006. Third annual report. Andhra Pradesh Water Management Project, ANGRAU University, Bapatla, India.

[26] Anjaneyulu, Y. 2003. Assessment of environmental quality of Kolleru Lake and strategic management plans. In: MK Durga Prasad and Y Anjaneyulu (Eds). Lake Kolleru - Environmental status (past and present). BS Publications, Hyderabad, 3-15.

[27] Anjaneyulu, Y, and Durga Prasad, MK. 2003. Lake Kolleru: environmental status (past and present). BS Publications, Hyderabad.

[28] Argent, RM, and Grayson, RB. 2003. A modelling shell for participatory assessment and management of natural resources. *Environmental Modelling & Software*, **18**, 541-551.

[29] Asian Development Bank. 2001. Project completion report on the Red River delta water resources sector project (loan no. 1344-VIE(SF)) in Vietnam. Asian Development Bank,, Ministry of Agriculture and Rural Development, Hanoi, Vietnam.

[30] Ayele, S, and Wield, D. 2005. Science and technology capacity building and partnership in African agriculture: perspectives on Mali and Egypt. *Journal of International Development*, **17**, 631-646, DOI: 10.1002/jid.1228.

[31] Ayers, RS, and Westcot, DW. 1985. *Water quality for agriculture.* Food and Agriculture Organization of the United Nations, Rome.

[32] Backeberg, GR. 2006. Reform of user charges, market pricing and management of water: problem or opportunity for irrigated agriculture. *Irrigation and Drainage*, **55**, 1-12.

[33] Badruddin, M. 2002. Drainage water reuse and disposal: a case study on Pakistan. In: KK Tanji and NC Kielen (Eds). Agricultural drainage water management in arid and semi-arid areas, Irrigation and Drainage Paper no. 61, Food and Agriculture Organization of the United Nations, Rome, 116-118.

[34] Bakker, N, Chu Tuan Dat, Smidt, P, and Steley, C. 2003. Developing a basin framework for prioritizing investments in water resources infrastructure in Vietnam's Red River Basin. Proceedings 9th International Drainage Workshop, Wageningen, 11.

[35] Berkhoff, K. 2007. Groundwater vulnerability assessment to assist the measurement planning of the water framework directive - a practical approach with stakeholders. *Hydrol. Earth Syst. Sci. Discuss.*, **4** (1133-1151).

[36] Bhutta, MH, Niazi, MFK, and Ahmed, N. 2003. Environmental impact of disposal of drainage effluent to evaporation ponds. Paper No 064, 9th International Drainage Workshop, September 10 – 13, 2003, Utrecht, The Netherlands.

[37] Bhutta, MH, and Smedema, LK. 2007. One hundred years of waterlogging and salinity control in the Indus Valley, Pakistan: a historical review. *Irrigation and Drainage*, **56**, S81 - S90. DOI: 10.1002/ird.333.

[38] Bhutta, MN, van der Sluis, TA, and Wolters, W. 1995. Review of pipe drainage projects in Pakistan. Proceedings of the National Workshop on drainage system performance in the Indus Plain and future strategies, Tando Jam, Pakistan, 10-18.

[39] Boonstra, J, and Bhutta, MH. 1996. Groundwater recharge in irrigated agriculture: the theory and practice of inverse modelling. *Journal of Hydrology*, **174**, 357-374.

[40] Bos, MG. 2006. Basics of groundwater flow. In: HP Ritzema (Ed). Drainage principles and applications, 3rd edition. ILRI Publication 16, Alterra-ILRI, Wageningen.

[41] Bos, MG, and Boers, TM. 2006. Land drainage: why and how? In: HP Ritzema (Ed). Drainage Principles and Applications, 3rd edition, ILRI Publication 16, Third Edition, Alterra-ILRI, Wageningen University and Research Centre, Wageningen, The Netherlands, 23-32.

[42] Bos, MG, and Boers, TM. 2007. Land drainage: Why and How? In: HP Ritzema (Ed). Drainage Principles and Applications, 3rd edition, ILRI Publication 16, Third Edition, Alterra-ILRI, Wageningen University and Research Centre, Wageningen, The Netherlands, 23-32.

[43] Bos, MG, and Wolters, W. 2006. Influences of irrigation on drainage. In: HP Ritzema (Ed). Drainage principles and applications, ILRI Publication 16, 3rd Edition, Alterra-ILRI, Wageningen, 513-532.

[44] Broughton, RS, and Fouss, JL. 1999. Subsurface drainage installation machinery and methods. In: RW Skaggs and J van Schilfgaarde (Eds). Agricultural drainage, American Society of Agronomy, 963-1003.

[45] Burt, CM, and Styles, SW. 1999. Modern water control and management practices in irrigation - impact on performance. Water reports no. 19, International Programme for Technology and Research in Irrigation and Drainage, The World Bank and Food and Agriculture Organization of the United Nations, Rome.

[46] Canter, LW. 1996. Environmental impact assessment. McGraw-Hill, USA.

[47] Capacity Building in the Water Resources Sector Project. 1999. Review of design process, criteria and standards in the water sector of Vietnam. Technical Report,

Capacity Building in the Water Resources Sector Project, TA No. 2233-VIE (TA-2), Ministry of Agriculture and Rural Development, Hanoi.

[48] Castelletti, R, and Soncini-Sessa, R. 2006. A procedural approach to strengthening integration and participation in water resources planning. *Environmental Modelling & Software*, **21** (1455-1470).

[49] Cavelaars, JC. 1985. Survey and design for subsurface drainage in Egypt. Advisory Panel For Land Drainage in Egypt, Cairo.

[50] Cavelaars, JC, Vlotman, WF, and Spoor, G. 2006. Subsurface drainage systems. In: HP Ritzema (Ed). Drainage principles and applications, 3 ed. , Alterra-ILRI, Wageningen, 827-930.

[51] Cavelaars, JC, Vlotman, WF, and Spoor, G. 2007. Subsurface drainage systems. In: HP Ritzema (Ed). Drainage Principles and applications, 3 ed. , Alterra-ILRI, Wageningen, 827-930.

[52] Chapagain, AK, Hoekstra, AY, Savenije, HHG, and Gautam, R. 2006. The water footprint of cotton consumption: an assessment of the impact of worldwide consumption of cotton products on the water resources in the cotton producing countries. *Ecological Economics*, **60**, 186-203.

[53] Croon, FW. 1997. Drainage of heavy clay soils in Egypt. Consultancy report, Drainage Research Programme Project (DRP), Drainage Research Institute, El Kanater, Egypt.

[54] Croon, FW. 1997. Institutional aspects of drainage implementation. 7th International Drainage Workshop "*Drainage for the 21st century*", 17-21 November 1997, Malaysian National Committee on Irrigation and Drainage, Penang.

[55] d'Aquino, P, Le Page, C, Bousquet, F, and Bah, A. 2002. A novel mediating participatory modelling: the 'self-design' process to *accompany* collective decision making. *Int. J. Agricultural Resources*, **2**, 59-74.

[56] Datta, KK, De Jong, C, and O.P., S. 2000. Reclaiming salt-affected land through drainage in Haryana, India: a financial analysis. *Agricultural Water Management*, **46**, 55-71.

[57] De Ridder, NA, and Boonstra, J. 2006. Analysis of water balances. In: HP Ritzema (Ed). Drainage principles and applications, 3rd edition. ILRI Publication 16, Alterra-ILRI, Wageningen, 601-633.

[58] Dolfing, B, and Snellen, WB. 1999. Sustainability of Dutch water boards: appropriate design characteristics for self-governing water management organisations. ILRI Special Report, ILRI, Wageningen.

[59] Doorenbos, J, and Kassam, AH. 1979. *Yield response to water*. Food and Agriculture Organization of the United Nation, Rome.

[60] Drainage Research Institute. 1992. Subsurface drainage system design for Haress pilot area. Technical Report no.70, Drainage Research Institute, Delta Barrages, Cairo Egypt.

[61] Drainage Research Institute. 1997. Drainage water irrigation project (DWIP) - Final Report. Drainage Research Institute, Cairo.

[62] Drainage Research Institute. 2000. Farmer participation in drainage management. Technical report 108, Drainage Research Project II, Drainage Research Institute, NWRC, Delta Barrages, Cairo.

[63] Drainage Research Project. 2001. Drainage research project I & II, final report, Dec 1994 – June 2001. Drainage Research Project, Drainage Research Institute, Kanater, Cairo.

[64] Drainage Technology and Pilot Areas Project. 1993. Final report - drainage technology and pilot areas project. Drainage Research Institute, Cairo.

[65] Durga Prasad, MK, and Padmavathi, P. 2003. Lake Kolleru - the biggest freshwater wetland ecosystem in South India: biodiversity and status. In: MK Durga Prasad and Y Anjaneyulu (Eds). Lake Kolleru - Environmental status (past and present). BS Publication, Hyderabad, 143-157.

[66] El-Atfy, HE, Abdel-Alim, MQ, and Ritzema, HP. 1991. A modified layout of the subsurface drainage system for rice areas in the Nile Delta, Egypt. *Agricultural Water Management*, **19**, 289-302.

[67] El-Atfy, HE, Abdel Alim, MQ, and Ritzema, HP. 1990. Experiences with a drainage system for rice areas in Egypt. Symposium on Land Drainage for Salinity Control in Arid and Semi-Arid Regions, Cairo, 129-141.

[68] El-Atfy, HE, El-Gamal, H, and van Mourik, E. 1991. Discharge rates, salinities, and the performance of subsurface collector drains in Egypt. *Irrigation and Drainage Systems*, **5**, 325-338.

[69] El-Atfy, HE, Wahid El-Din, O, El-Gammal, H, and Ritzema, HP. 1990. Hydraulic performance of subsurface collector drains in Egypt. . Symposium on Land Drainage for Salinity Control in Arid and Semi-Arid Regions, Cairo, 393-404.

[70] El-Guindi, S, and Amer, MH. 1989. Establishment of the advisory panel on land drainage. In: MH Amer and NA De Ridder (Eds). Land Drainage in Egypt, Drainage Research Institute, Cairo, 95-101.

[71] El-Guindi, S, and Risseeuw, IA. 1987. *Research on Water Management of Rice Fields in the Nile Delta, Egypt.* ILRI Publication 41, International Institute for Land Reclamation and Improvement, Wageningen.

[72] El-Quosy, DE. 1989. Drainage water re-use projects in the Nile delta: the past, the present and the future. In: MH Amer and NA De Ridder (Eds). Land drainage in Egypt, Drainage Research Institute, Cairo, 163-175.

[73] Fahlbusch, H, Schultz, B, and Thatte, CD. 2004. The Indus basin: history of irrigation, drainage and flood management. International Commission on Irrigation and Drainage, New Delhi.

[74] FAO. 1979. *Yield response to water.* Food and Agriculture Organization of the United Nation, Rome.

[75] Fathi, M, and Hamza, AM. 2000. Development of land drainage in Egypt and the role of the Egyptian Public Authority for Drainage Projects. In: HJ Nijland (Ed). Drainage along the River Nile, Ministry of Public Works and Water Resources, Egypt, Ministry of Transport, Public Works and Water Management, Directorate-General of Public Works and Water Management, The Netherlands, Lelystad, 21-51.

[76] Feddes, RA, and Lenselink, KJ. 2006. Evaporation. In: HP Ritzema (Ed). Drainage principles and applications, 3rd edition. ILRI Publication 16, Alterra-ILRI, Wageningen, 145-173.

[77] Fontenelle, JP. 1999. The response of farmers to political change: decentralization of irrigation in the Red River Delta, Vietnam. Liquid Gold paper no. 5, International Institute for Land Reclamation and Improvement, Wageningen.

[78] Fontenelle, JP. 2000. Local institutional innovation in Red River delta irrigation management. Irrigation and Water Engineering, Wageningen University.

[79] Framji, KK, Garg, BC, and Kaushish, SP. 1984. *Design practices of open drainage channels in an agricultural land drainage system: a worldwide survey.* International Commission on Irrigation and Drainage, New Delhi.

[80] Franks, T, Garcés-Restrepo, C, and Pututhena, F. 2008. Developing capacity for agricultural water management: current practice and future directions. *Irrigation and Drainage*, **57**, 255-267, DOI: 10.1002/ird.433.

[81] Fujisaka, s. 1997. Research: help or hindrance to good farmers in high risk systems? *Agricultural systems*, **54** (2), 137-152.

[82] Gilbert, E, Posner, J, and Sumberg, J. 1990. Farming systems research within a small research system: a search for appropriate models. *Agricultural systems*, **3**, 327-346.

[83] Global Water Partnership. 2003. Sharing knowledge for equitable, efficient and sustainable water resources. Tool box integrated water resources management, Global Water Partnership, Stockholm, Sweden.

[84] Goosen, H. 2007 ? Spatial water management (in Dutch). Vrije Universitiet, Amsterdam.

[85] Gopalakrishnan, M. 2008. Egypt's paradigm in integrated water resources management (IWRM). *Irrigation and Drainage*, **56**, 487-488, DOI: 10.1002/ird.331.

[86] Gopalakrishnan, M, and Kulkarni, SA. 2007. Agricultural land drainage in India. *Irrigation and Drainage*, **56**, S59-S67. DOI: 10.1002/ird.368.

[87] Goss, D. 2004. Fieldbook for participatory learning and action. DWC and InWENT, Hanoi Department of Culture and Information, Hanoi.

[88] Government of Andhra Pradesh. 1999. Declaration of areas for Kolleru wildlife sanctuary. F Environment, Science and Technology Department, Government of Andhra Pradesh, ed., The Andhra Pradesh Gazette, 6.

[89] Government of India. 2005. Annual report 2003-2004. Government of India, Ministry of Water Resources, New Delhi.

[90] Gupta, SK. 2002. A century of subsurface drainage research in India. *Irrigation and Drainage Systems*, **16**, 69-84.

[91] Gupta, SK, and Gupta, IC. 1997. *Crop production in waterlogged saline soils.* Pawan Kumar Scientific Publishers, Jodhpur, India.

[92] Hamilton, AJ, Stagnitti, F, Xiong, X, Kreidl, SL, Benke, KK, and Maher, P. 2007. Wastewater irrigation: the state of play. *Vadose Zone Journal*, **6**, 823-840 (DOI: 10.2136/vzj2007.0026).

[93] HarmoniCOP. 2005. Learning together to manage together - improving participation in water management. Harmonising collaborative planning, D Ridder, E Mostert, and HA Wolters, eds., Druckhaus Bergman, 99.

[94] Hoekstra, AY. 2005. Generalisme als specialisme - waterbeheer in context van duurzame ontwikkeling, globalisering, onzekerheden en risico's (in Dutch). University of Twente, Enschede.

[95] Hoekstra, AY. 2008. Water neutral: reducing and offsetting the impacts of water footprints. Value of water research report series no. 28, UNESCO-IHE, Delft.

[96] Hoekstra, AY, and Chapagain, AK. 2007. Water footprints of nations: water use by people as a function of their consumption pattern. *Water Resources Management*, **21**, 35-48.

[97] Huamanchumo, J, Peña, Y, Silva, L, and Hendriks, J. 2008. Developing capacity in water users organizations: the case of Peru. *Irrigation and Drainage*, **57**, 300-310, DOI: 10.1002/ird.432.

[98] Hussein, MH, and Hoogenboom, PJ. 1999. Trenchless drainage technique and its economy under Egyptian conditions. ILRI-IPTRID Workshop, Cairo, 4.

[99] Hvidt, M. 1996. Improving irrigation system performance in Egypt: first experiences with the WUA approach. *Water Resources Development*, **12** (3), 261-276.

[100] Indian National Committee on Irrigation and Drainage. 2000. New Delhi declaration. 8th ICID International Drainage Workshop with the theme "Role of Drainage and Challenges in 21st Century" New Delhi.

[101] Indo-Dutch Network Project. 2002. Computer modelling in irrigation and drainage. Indo-Dutch Network Project, Central Soil Salinity Research Institute, Karnal, India.

[102] Indo-Dutch Network Project. 2002. Human resources development and establishment of a training centre. Indo-Dutch Network Project, Central Soil Salinity Research Institute, Karnal and Alterra-ILRI, Wageningen.

[103] Indo-Dutch Network Project. 2002. Recommendations on waterlogging and salinity control based on pilot area drainage research. Indo-Dutch Network Project, Central Soil Salinity Research Institute, Karnal, India.

[104] International Assessment of Agricultural Knowledge Science and Technology for Development. 2008. Synthesis report of the intergovernmental plenary session in Johannesburg, South Africa in April, 2008. International Assessment of Agricultural Knowledge, Science and Technology for Development.

[105] International Commission on Irrigation and Drainage. 2003. Important data of ICID member countries. database on website: www.icid.org., International Commission on Irrigation and Drainage

[106] International Commission on Irrigation and Drainage. 2008. Water resources and irrigation development in Pakistan. ICID Newsletter, 2008/3, 3.

[107] International Conference on Water and the Environment. 1992. The Dublin Statement on Water and Sustainable Development. Dublin, 6.

[108] International Institute for Land Reclamation and Improvement. 1992-1999. Annual Report. International Institute for Land Reclamation and Improvement, International Institute for Land Reclamation and Improvement/ILRI, Wageningen.

[109] International Programme for Technology and Research in Irrigation and Drainage Secretariat. 2007. Synthesis paper. Proceedings Workshop on Monitoring and Evaluation of Capacity Development Strategies in Agricultural Water Management, Kuala Lumpur, 73-85.

[110] Jansen, HC, Bhutta, MN, Javed, I, and Wolters, W. 2006. Groundwater modelling to assess the effect of interceptor drainage and lining – example of model application in the Fordwah Eastern Sadiqia project, Pakistan. *Irrigation and Drainage Systems*, **20**, 23-40.

[111] Jaspers, AMJ, and Schrevel, A. 1999. Capacity building in irrigated agriculture. In: A Schrevel (Ed). Water and food security in (semi-) arid areas, Wageningen, 147-156.

[112] Javed, I, Boonstra, J, and Hafeez, A. 1999. Groundwater study at Fordwah Eastern Sadiqia (South) Project, Bahawalnager. IWASRI Publication 221, International Waterlogging and Salinity Research Institute, Lahore.

[113] Javed, I, and Hafeez, A. 2004. Disposal of drainage effluent in evaporation ponds of Pakistan. 2nd Asia Regional Conference, March 2004, Beijing.

[114] Jeffrey, P, and Russell, S. 2007. *Participative planning for water reuse projects, a handbook of principles, tools & guidance* Aquarec project, http://www.aquarec.org/.

[115] Jonsson, A, Andersson, L, Alkan-Olsson, J, and Arheimer, B. 2007. How participatory can participatory modeling be? Degrees of influence of stakeholder and expert perspectives in six dimensions of participatory modeling. *Water Science & Technology*, **56** (1), 207-214.

[116] Joosten, H, and Clark, D. 2002. *Wise use of mires and peatlands – background and principles including a framework for decision-making.* International Mire Conservation Group and International Peat Society.

[117] Jurriens, M, Mollinga, PP, and Wester, P. 1996. Scarcity by design - protective irrigation in India and Pakistan. Special Report., International Institute for Land Reclamation and Improvement and Wageningen Agricultural University, Wageningen.

[118] Kahlown, MA, Marri, MK, and Azam, M. in press. Design, construction and performance evaluation of small tile drainage systems in the Indus Basin. *Irrigation and Drainage* DOI: 10.1002/ird.342.

[119] Kay, M, and Terwisscha van Scheltinga, C. 2004. Towards sustainable irrigation and drainage through capacity building. 9th International Drainage Workshop, September 10-13, 2003, Utrecht, The Netherlands, 16.

[120] Khan, MA, Bhutta, MH, and Wolters, W. 1997. Performance of surface drains in Pakistan. Proceedings of the 7th International Drainage Workshop: Drainage for the 21st Century, Penang, 2-10.

[121] Kielen, NC. 2002. Drainage water reuse and disposal: a case study from the Nile Delta, Egypt. In: KK Tanji and NC Kielen (Eds). Agricultural drainage water management in arid and semi-arid areas, Irrigation and Drainage Paper no. 61, Food and Agriculture Organization of the United Nations, Rome, 113-114.

[122] Kishwar, I, and Donaldson, AP. 1997. Baseline socio-economic survey joint Satiana pilot project. IWASRI Publication 166., International Waterlogging and Salinity Research Institute, Lahore.

[123] Knops, JAC, Alam, MM, and Ilyas, M. 1999. Proceedings of the second national experts consultation on farmers' participation in drainage. Lahore.

[124] Knops, JAC, Bhutta, MN, Malik, H, and Wolters, W. 1996. Towards improved drainage performance in Pakistan. Transactions Workshop on the Evaluation of Performance of Subsurface Drainage Systems, 16th Congress on Irrigation and Drainage, ICID, Cairo, 127-138.

[125] Knops, JAC, and Siddiq, M. 1997. Proceedings of the national experts consultation on farmers' participation in drainage. Lahore.

[126] Kumbhare, PS, and Ritzema, HP. 2000. Need, selection and design of synthetic envelopes for subsurface drainage in clay and sandy loam soils of India. 8th ICID International Drainage Workshop, New Delhi, India, VI163-VI175.

[127] La Grusse, P, Belhouchette, H, Le Bars, M, Carmona, G, and Attonaty, JM. 2006. Participative modelling to help collective decision-making in water allocation and nitrogen pollution: application to the case of the Aveyron-Lère Basin. *Int. J. Agricultural Resources*, **5** (2/3), 247-271.

[128] Leathes, W, Knox, JW, Kay, MG, Trawick, P, and Rodriguez-Diaz, JA. 2008. Developing UK farmers' institutional capacity to defend their water rights and effectively manage limited water resources. *Irrigation and Drainage*, **57**, 322-331, DOI: 10.1002/ird.436.

[129] Lee, P. 2007. Top 10 irrigation technologies. ICID Newsletter, New Delhi.

[130] Linh, NV. 2001. Agricultural innovation: multi ground for technology policies in the Red River delta of Vietnam. PhD dissertation no. 2956, Wageningen University Wageningen.

[131] Loucks, DP. 2006. Modeling and managing the interactions between hydrology, ecology and economics. *Journal of Hydrology*, **328**, 408-416.

[132] Luijendijk, J, and Lincklean Arriëns, W. 2007. Water knowledge networking: partnering for better results. Water for a changing world - enhancing local knowledge and capacity., Delft, June 13-15., 25.

[133] Luijendijk, J, and Mejia-Velez, D. 2005. Knowledge networks for capacity building: a tool for achieving MDGs? Design and Implementation of Capacity Development Strategies, Beijing, China, 14 September 2005, 113-131.

[134] Malano, HM, and Van Hofwegen, PJM. 1999. *Management of irrigation and drainage systems - a service approach.* Balkema, Rotterdam.

[135] Máñez, M, Froebrich, J, Ferrand, N, and Siva, A. 2007. Participatory dam systems modelling: a case study of transboundary Guadianan River in the Iberian Peninsula. *Water Science & Technology*, **56** (4), 145-156.

[136] Manjunatha, MV, Oosterbaan, RJ, Gupta, SK, Rajkumar, H, and Jansen, H. 2004. Performance of subsurface drains for reclaiming waterlogged saline lands under rolling topography in Tungabhadra irrigation project in India. *Agricultural Water Management*, **69**, 69-82.

[137] Mann, AA, Saif-ur-Rehman, and Iqbal, M. 1997. Impact assessment of a sub-surface drainage system in the Fourth Drainage Project of Faisalabad. IWASRI Internal Report 97/21.1, International Waterlogging and Salinity Research Institute, Lahore.

[138] Mathew, EK, Nair, M, Raju, TD, and Jaikumaran, U. 2003. *Drainage digest - a report based on two decades (1981-2002) of research under AICRP (Drainage) at the Karumady Centre, Kerala, India.* Kerala Agricultural University, Thrissur, Kerala, India.

[139] McCready, W. 1987. Left bank outfall drain in Pakistan. ICID Bulletin 36, International Commission on Irrigation and Drainage, 15-19.

[140] Meinzen-Dick, RS, and Van der Hoek, W. 2001. Multiple uses of water in irrigated areas. *Irrigation and Drainage Systems*, **15**, 93-98.

[141] Menshawy, R, Fokkens, B, Ali, AM, and Nijland, HJ. 2000. Operational research in drainage. In: HJ Nijland (Ed). Drainage along the River Nile, RIZA Nota nr. 2000.052, Ministry of Public Works and Water Resources, Egypt, Ministry of Transport, Public Works and Water Management, Directorate-General of Public Works and Water Management, The Netherlands, 167-178.

[142] Menshawy, R, Penninkhof, J, Shaheen, OS, and Visser, HJP. 2000. Capacity and efficiency of drainage machines in Egypt. In: HJ Nijland (Ed). Drainage along the River Nile, RIZA Nota nr. 2000.052, Ministry of Public Works and Water Resources, Egypt, Ministry of Transport, Public Works and Water Management, Directorate-General of Public Works and Water Management, The Netherlands, 179-187.

[143] Merrey, DJ. 2008. Will future water professionals sink under received wisdom, or swim to a new paradigm? The water professional of tomorrow, on the occasion of the World Water Day, Netherlands National ICID Committee, Wageningen.

[144] Mohtadullah, K, Bhutta, MN, Wolters, W, and Alam, MM. 1997. Benefits of linking research with design and implementation of drainage projects. Proceedings of the 7th International Drainage Workshop: Drainage for the 21st Century, Penang, T7:1-15.

[145] Molden, D. 2007. Water for food, water for life: a comprehensive assessment of water management in agriculture. London: Earthscan and Colombo: International Water Management Institute.

[146] Mustajoki, J, Hämäläinen, RP, and Marttunen, M. 2004. Participatory multicriteria decision analysis with Web-HIPRE: a case of lake regulation policy. *Environmental Modeling & Software*, **19**, 537-547.

[147] MX.Systems, Sa. 2004. Duflow modelling studio – user's guide, version 2.7. Stowa, Utrecht

[148] Nageswara Rao, K, Murali Krishna, G, and Hema Mallini, B. 2004. Kolleru Lake is vanishing – a revelation through digital processing of IRS-1D LISS-III sensor data. *Current Science*, **86** (9), 1312-1316.

[149] Nijland H.J. (Ed.). 2000. *Drainage along the River Nile.* Ministry of Public Works and Water Resources, Egypt, Ministry of Transport, Public Works and Water Management, Directorate-General of Public Works and Water Management, The Netherlands.

[150] Nijland, HJ, Croon, FW, and Ritzema, HP. 2005. *Subsurface drainage practices: guidelines for the implementation, operation and maintenance of subsurface pipe drainage systems. ILRI Publication 60.* Alterra, Wageningen University and Research Centre, Wageningen.

[151] Nijland, HJ, and El-Guindi, S. 1984. Crop yield, groundwatertable depth, and soil salinity in the Nile Delta, Egypt. Annual Report 1983, Wageningen, 19-29.

[152] Nijland, HJ, Salman, AF, and Youssef, YM. 2000. Drainage technology developments in the large-scale execution of drainage projects in Egypt. In: HJ Nijland (Ed). Drainage along the River Nile, RIZA Nota nr. 2000.052, Drainage Executive Management Project, Ministry of Public Works and Water Resources, Egypt, Ministry of Transport, Public Works and Water Management, Directorate-General of Public Works and Water Management, The Netherlands., 253-263.

[153] Nonaka, I, and Takeuchi, H. 1995. *The knowledge-creating company; how Japanese companies create the dynamics of innovation.* Oxford University Press, New York.

[154] Ochs, WJ, and Bishay, GB. 1992. Drainage guidelines. Technical paper no. 195, World Bank, Washington D.C.

[155] Oosterbaan, RJ. 1987. Report of a consultancy assignment to the pilot areas and drainage technology project of the Drainage Research Institute, March 1987. International Institute for Land Reclamation and Improvement, Wageningen, The Netherlands.

[156] Oosterbaan, RJ. 1988. Agricultural criteria for subsurface drainage: a system analysis. *Agricultural Water Management*, **14**, 79-90.

[157] Oosterbaan, RJ. 2000. *SALTMOD – Description of principles, user manual and examples of application.* International Institute for Land Reclamation and Improvement, Wageningen.

[158] Oosterbaan, RJ, and Abu Senna, M. 1990. Drainage and salinity predictions in the Nile Delta using SALTMOD. Symposium on Land Drainage for Salinity Control in Arid and Semi-Arid Regions, Cairo, 274-286.

[159] Pangare, G, Pangare, V, and Das, B. 2006. *Springs of life – India's water resources.* Academic Foundation, New Delhi, India.

[160] Pant, N. 2000. Drainage institutions in Western Europe: England, the Netherlands, France and Germany. The World Bank, Washington.

[161] Parikh, MM, Lad, AN, Shrivastava, PK, Patel, AM, and Raman, S. 1999. Pre-drainage investigations in Segwa pilot area. Tech. Rep 1, Indo-Dutch Network Project, Soil and Water Management Research Unit, Gujarat Agricultural University, Navsari.

[162] Parikh, MM, Patel, BR, Srivastava, PK, Lad, AN, Patel, AM, and Raman, S. 1999. Drainage needs and research achievement in South Gujarat. Soil and Water Management Research Unit. Water Management Research in Gujarat, Gujarat Agricultural University, Navsari, Gujarat, India, 217-227.

[163] Pavelis, GA. 1987. Farm drainage in the United States: history, status and prospects. US Department of Agriculture, Economic Resources Service, Washington, 170.

[164] Pearce, G, and Dennecke, HW. 2001. Drainage and sustainability. Issue Paper no. 3, IPTRID/FAO, Rome.

[165] Pilarczyk, KW, and Nuoi, NS. 2005. Experience and practices on flood control in Vietnam. *Water International*, **30** (1), 114-122.

[166] Pilot Areas and Drainage Technology Project. 1984. Monitoring the performance of a drainage system with modified layout and comparison with a conventional system - summer season 1983. Technical Report no. 27, Pilot Areas and Drainage Technology Project, Drainage Research Institute, Cairo.

[167] Pilot Areas and Drainage Technology Project. 1985. Monitoring the performance of a drainage system with modified layout and comparison with a conventional system - summer season 1984. Technical Report no. 30, Pilot Areas and Drainage Technology Project, Drainage Research Institute, Cairo.

[168] Pilot Areas and Drainage Technology Project. 1986. Design and monitoring of modified drainage systems - summer 1985. Technical Report no. 49, Pilot Areas and Drainage Technology Project, Drainage Research Institute, Cairo.

[169] Pilot Areas and Drainage Technology Project. 1986. Water management in rice fields under conventional and modified concepts - summer season 1986. Technical Report no. 54, Pilot Areas and Drainage Technology Project, Drainage Research Institute, Cairo.

[170] Pilot Areas and Drainage Technology Project. 1986. Water management in rice fields under conventional and modified drainage concepts. Technical Report no. 48, Pilot Areas and Drainage Technology Project, Drainage Research Institute, Cairo.

[171] Pilot Areas and Drainage Technology Project. 1987. Mashtul pilot area: physical description. Pilot Areas and Drainage Technology Project, Technical Report no.57, Drainage Research Institute, Giza, Egypt.

[172] Pilot Areas and Drainage Technology Project. 1987. Monitoring the hydraulic performance of collector drains in Namul and Roda areas - summer 1986. Technical Report no. 55, Pilot Areas and Drainage Technology Project, Drainage Research Institute, Cairo.

[173] Pilot Areas and Drainage Technology Project. 1987-89. Drainage criteria study at Mashtul pilot area. Pilot Areas and Drainage Technology Project, Technical Report No 59, part I - VII, Drainage Research Institute, Giza, Egypt.

[174] Pilot Areas and Drainage Technology Project. 1989. Monitoring of conventional and modified drainage systems - summer 1988. Technical Report no. 63, Pilot Areas and Drainage Technology Project, Drainage Research Institute, Cairo.

[175] Pilot Areas and Drainage Technology Project. 1990. Drainage criteria study at Mashtul pilot area - final report. Technical Report no. 64, Pilot Areas and Drainage Technology Project, Drainage Research Institute, El-Quanater.

[176] Postel, S. 1999. *Pillar of sand - Can the irrigation miracle last?* Worldwatch Institute, New York.

[177] Qureshi, AS, Iqbal, M, Anwar, NA, Aslam, M, and R.M., C. 1997. Benefits of shallow drainage. Proceedings Seminar on-farm salinity, drainage, and reclamation. IWASRI Publication 179.

[178] Raats, PAC, and Feddes, RA. 2006. Contributions by Jans Wesseling, Jan van Schilfgaarde and Herman Bouwer to effective and responsible water management in agriculture. *Agricultural Water Management*, **86**, 9-29, DOI:10.1016/j.agwat.2006.10.014.

[179] Rafiq, M. 1998. Field testing and evaluation of envelope materials for pipe drains. Thesis for partial fulfilment of requirements for PhD degree, CEWRE, UET, Lahore.
[180] Rafiq, M, Bhutta, MN, Chaudhry, R, Khan, MA, and Subhani, KM. 2000. Farmers' participatory drainage in the Bahawalnagar pilot study area. IWASRI Publication 211, International Waterlogging and Salinity Research Institute, Lahore.
[181] Rajalahti, R, Woelcke, J, and Pehu, E. 2005. Developing research systems to support the changing agricultural sector. Agriculture and Rural Development Discussion paper 14, The World Bank, Washington.
[182] Rajasthan Agricultural Drainage Research Project. 1996. Analysis of subsurface drainage design criteria. Rajasthan Agricultural Drainage Research Project, Chambal Command Area Development Authority, Rajasthan, India, Nutech Photolithographers, New Delhi, India.
[183] Raju, KRK. 2006. Area of Lake Kolleru converted into aquaculture (data obtained from satellite images). SRKR Engineering College, Undi.
[184] Ramani, KV, and Anjaneyulu, Y. 2003. Pollution studies of Kolleru Lake by industrial, agricultural, domestic wastes and corrective management plans. In: MK Durga Prasad and Y Anjaneyulu (Eds). Lake Kolleru - Environmental status (past and present), BS Publication, Hyderabad, 143-157.
[185] Rao, KVGK. 1994. Drainage of salt affected soils. In: DLN Rao, NT Singh, RK Gupta, and NK Tyagi (Eds). Salinity management for sustainable agriculture – 25 years of research at CSSRI, Central Soil Salinity Research Institute, Karnal, India, 177-200.
[186] Rao, KVGK, Agarwal, MC, Singh, OP, and Oosterbaan, RJ. 1995. Reclamation and management of waterlogged saline soils, National Seminar Proceedings - summary and recommendations. iii-ix.
[187] Rao, KVGK, Sharma, SK, and Kumbhare, PS. 1995. Drainage requirements of alluvial soils of Haryana. In: KVGK Rao, MC Agarwal, OP Singh, and RJ Oosterbaan (Eds). National Seminar on the reclamation and management of waterlogged saline soils, Karnal, India, 36-49.
[188] Reddy, MS, and Char, NVV. 2004. Management of lakes in India., Ministry of Water Resources, Delhi, 20.
[189] Research Institute for Groundwater. 1988. Groundwater development in the Eastern Nile Delta. Technical Note 70-120-88-06, Research Institute for Groundwater, Cairo.
[190] Ridder, D, Mostert, E., Wolters, H.A. (Eds),. 2005. Learning together to manage together - improving participation in water management. Harmonising collaborative planning, Druckhaus Bergman, 99.
[191] Ritzema H.P. (Ed.). 1994. Drainage principles and applications. Second completely revised edition. ILRI Publication 16, International Institute for Land Reclamation and Improvement, Wageningen, 1125.
[192] Ritzema H.P. (Ed.). 2006. Drainage principles and applications, 3rd edition. ILRI Publication 16. ILRI Publication 16, Alterra-ILRI, Wageningen.
[193] Ritzema, HP. 1994. Subsurface flow to drains. In: HP Ritzema (Ed). Drainage principles and applications, ILRI Publication 16, International Institute for Land Reclamation and Improvement, Wageningen, 263-304.
[194] Ritzema, HP. 2007. Performance assessment of subsurface drainage systems - case studies from Egypt and Pakistan. Alterra, Wageningen.
[195] Ritzema, HP, and Abdel Alim, MQ. 1985. Testing a rod rotator unit for cleaning subsurface drainage systems. Technical Report no. 52, Drainage Research Institute, El Kanater, Cairo.

[196] Ritzema, HP, and Braun, HMH. 1994. Environmental aspects of drainage. In: HP Ritzema (Ed). Drainage Principles and Applications. ILRI Publication 16., International Institute for Land Reclamation and Improvement, Wageningen, 1041-1066.

[197] Ritzema, HP, Elbers, HAJ, and Jallow, BG. 1995. Rice cultivation in tidal areas in the Gambia. ILRI Annual Report 1995, International Institute for Land Reclamation and Improvement, Wageningen, 8-23.

[198] Ritzema, HP, Kselik, RAL, and Chanduvi, F. 1996. Drainage of irrigated lands. Irrigation water management training manual, no. 9, Food and Agriculture Organization of the United Nations, Rome, 74.

[199] Ritzema, HP, Murtedza, M, Page, S, Limin, S, and Wösten, JHM. 2006. Capacity building for sustainable management of peatlands in the humid tropics: from research to application. Workshop on Monitoring and Evaluation of Capacity Development Strategies in Agricultural Water Management, Kuala Lumpur, Malaysia, 73-85.

[200] Ritzema, HP, Mutalib Mat Hassan, A, and Moens, RP. 1998. A new approach to water management of tropical peatlands: a case study from Malaysia. *Irrigation and Drainage Systems*, **12**, 123-139.

[201] Ritzema, HP, Nijland, HJ, and Croon, FW. 2006. Subsurface drainage practices: from manual installation to large-scale implementation. *Agricultural Water Management*, **86**, 60-71. DOI:10.1016/j.agwat.2006.06.026.

[202] Ritzema, HP, Raman, S, Satyanarayana, TV, and Boonstra, J. 2003. Drainage protects irrigation investments. Indo-Dutch Network Project, Alterra-ILRI, Wageningen.

[203] Ritzema, HP, Satyanarayana, TV, Raman, S, and Boonstra, J. 2008. Subsurface drainage to combat waterlogging and salinity in irrigated lands in India: lessons learned in farmers' fields. *Agricultural Water Management* **95**, 179 - 189. DOI: 10.106/j.agwat.2007.09.012.

[204] Ritzema, HP, Thinh, LD, Anh, LQ, Hanh, DN, Chien, NV, Lan, TN, Kselik, RAL, and Kim, BT. 2008. Participatory research on the effectiveness of drainage in the Red River Delta, Vietnam. *Irrigation and Drainage Systems* **22**, 19–34. DOI: 10.1007/s10795-007-9028-0.

[205] Ritzema, HP, and Wolters, W. 2002. New frontiers in capacity building in drainage. Capacity building for drainage in North Africa, Cairo, Egypt, 10-14 March 2001, 29-38.

[206] Ritzema, HP, Wolters, W, Bhutta, MN, Gupta, SK, and Abdel-Dayem, S. 2007. The added value of research on drainage in irrigated agriculture. *Irrigation and Drainage*, **56**, S205 - S215. DOI: 10.1002/ird.337.

[207] Ritzema, HP, Wolters, W, and Terwisscha van Scheltinga, CTHM. 2008. Lessons learned with an integrated approach for capacity development in agricultural land drainage. *Irrigation and Drainage*, **57**, 354–365. DOI: 10.1002/ird.431.

[208] Ritzema, HP, and Wösten, JHM. 2003. Water management: the key for agricultural development of the lowland peat swamps of Borneo. International Workshop on Sustainable Development of Tidal Areas, International Commission on Irrigation and Drainage, Montreal, 101-113.

[209] Roest, K, Ritzema, HP, and Wolters, W. 2006. Drainage: the forgotten factor. Symposium on Irrigation Modernization: Constraints and Solutions, International Programme for Technology and Research in Irrigation and Drainage (IPTRID) of the Food and Agriculture Organization (FAO), Damascus, Syria.

[210] Roux, DJ, Rogers, KH, Biggs, HC, Ashton, PJ, and Sergeant, A. 2006. Bridging the science–management divide: moving from unidirectional knowledge transfer to knowledge interfacing and sharing. *Ecology and Society*, **11** (1), 4.

[211] Rudra, A. 1978. Organisation of agriculture for rural development: the Indian case *Camb. J. Econ.*, **2** (4), 318-406.

[212] Salman, M, and Garces, C. 2007. Egypt's experience in irrigation and drainage research uptake   - final report. IPTRID Secretariat, Food and Agriculture Organization of the UN, Rome.

[213] Satyanarayana, TV, and Boonstra, J. 2007. Subsurface drainage pilot areas experiences in three irrigated project commands of Andhra Pradesh in India. *Irrigation and Drainage*, **56**, S245 - S522. DOI: 10.1002/ird.365.

[214] Satyanarayana, TV, Lakshmi, GV, Hanumanthaiah, CV, Srinivasulu, A, and Ratnam, M. 2003. Feasible subsurface drainage strategies to combat waterlogging and salinity in irrigated agricultural lands in Andhra Pradesh. Indo-Dutch Network Bridging Project on Drainage and Water Management, Acharya N.G. Ranga, Agricultural University, Bapatla, Andhra Pradesh., India.

[215] Scheumann, W. 1997. Managing salinization, institutional analysis of public irrigation systems. University of Berlin, Berlin.

[216] Schofield N. 2005. Land & Water Australia's Portfolio: Return on Investment and Evaluation Case Studies. Land & Water Australia, Agtrans Research, Canberra.

[217] Schultz, B. 1990. Guidelines on the construction of horizontal subsurface drainage systems. International Commission on Irrigation and Drainage, New Delhi, 236.

[218] Schultz, B, Thatte, CD, and Labhsetwar, VK. 2005. Irrigation and drainage: main contributors to global food production. *Irrigation and Drainage*, **54**, 263 - 278. DOI: 10.1002/ird.170.

[219] Schultz, B, Zimmer, D, and Vlotman, WF. 2007. Drainage under increasing and changing requirements. *Irrigation and Drainage*, **56**, S3 - S22. DOI: 10.1002/ird.372.

[220] Schwab, GO, and Fouss, JL. 1999. Drainage materials. In: RW Skaggs and J van Schilfgaarde (Eds). Agricultural drainage, American Society of Agronomy, 911-962.

[221] Shafiq-ur-Rehman, Broughton, RS, Bonnell, RB, and Alam, MM. 1996. Laboratory tests of fabric-sand envelopes for pipe drains. . Transactions Workshop on the Evaluation of Performance of Subsurface Drainage Systems. 16th Congress on Irrigation and Drainage, ICID, Cairo, 153-164.

[222] Sharma, DP, and Gupta, SK. 2006. Subsurface drainage for reversing degradation of waterlogged saline lands. *Land degradation and development*, **17**, 605 - 615. DOI: 10.1002/ldr.737.

[223] Sharma, DP, and Tyagi, NK. 2004. On-farm management of saline drainage water in arid and semi-arid regions. *Irrigation and Drainage*, **53**, 87-103.

[224] Shivaji Rao, T. 2003. Conflict between development and environment of Kolleru Lake area. In: Y Anjaneyulu and MK Durga Prasad (Eds). Lake Kolleru: environmental status (past and present), BS Publications, Hyderabad, 168-196.

[225] Shrivastava, PK, Pater, BR, Patel, AM, and Raman, S. 2001. Salt and water balance study at Segwa pilot area. Tech. Rep. 8, Indo-Dutch Network Project, Soil and Water Management Research Unit, Gujarat Agricultural University, Navsari.

[226] Singh, M, Bhattacharya, AK, Nair, TVR, and Singh, AK. 2002. Nitrogen loss through subsurface drainage effluent in coastal rice field from India. *Agricultural Water Management*, **52**, 249-260.

[227] Skaggs, RW, and Chescheir, GM. 2003. Effects of subsurface drain depth on nitrogen losses from drained lands. *Transactions of the American Society of Agricultural Engineers*, **46** (2), 237-244.

[228] Skaggs, RW, and Van Schilfgaarde, J. 1999. *Agricultural drainage*. No. 38 series of Agronomy, American Society of Agronomy, Madison.

[229] Skaggs, RW, and Van Schilfgaarde, J. 1999. Introduction. In: RW Skaggs and J Van Schilfgaarde (Eds). Agricultural drainage, American Society of Agronomy, 3-10.

[230] Slootweg, R, Hoevenaars, J, and Abdel-Dayem, S. 2007. DRAINFRAME as a tool for integrated strategic environmental assessment: lessons from practice. *Irrigation and Drainage*, **56**, S191-S203.

[231] Smedema, LK. 1997. Drainage development and research and training needs in the semi-humid and humid tropics. Drainage for the 21st century, Proceedings 7th International Drainage Workshop, Penang, K5-11.

[232] Smedema, LK. 2000. Global drainage needs and challenges. 8th ICID International Drainage Workshop, New Delhi.

[233] Smedema, LK. 2007. Nine international drainage workshops: history, objectives, contents and reflections on significance and impacts. *Irrigation and Drainage*, **56**, S23-S34 (DOI: 10.1002/ird.341).

[234] Smedema, LK. 2007. Revisiting currently applied pipe drain depths for waterlogging and salinity control of irrigated lands in the (semi) arid zone. *Irrigation and Drainage*, **56** (379-387).

[235] Smedema, LK. 2008. Drainage development: driving forces, government policies, market assessment and country cases (unpublished work). 10.

[236] Smedema, LK, and Shiati, K. 2002. Irrigation and salinity: a perspective review of the salinity hazards of irrigation development in the arid zone. *Irrigation and Drainage Systems*, **16**, 161-174.

[237] Smedema, LK, Wolters, W, and Hoogenboom, PJ. 1992. Reuse simulation in irrigated river basin. *Journal of Irrigation and Drainage Engineering*, **118** (6), 841-851.

[238] Snellen, WB. 1997. Towards integration of irrigation and drainage management. Jubilee symposium at the occasion of the fortieth anniversary of ILRI and the thirty-fifth anniversary of the ICLD, Wageningen, ILRI Special Report, 172.

[239] Snellen, WB. unpublished. Agriculture: the Indian irrigation sector's forgotten objective. Alterra, Wageningen, 18.

[240] Soil and Water Management Research Unit. 2003. Subsurface drainage strategies to combat waterlogging and salinity in canal commands of Gujarat. Indo-Dutch Network Project, Gujarat Agricultural University, Navsari, Gujarat, India.

[241] Sreenivas, C. 2006. Collaborative research report on Kolleru Lake and Upputeru Estuary rehabilitation study., Acharya N.G. Ranga Agricultural University, Undi.

[242] Sreenivas, C, and Konda Reddy, C. 2008. Salinity-sodicity relationships of the Kalipatnam drainage pilot area, Godavari Western Delta, India. *Irrigation and Drainage* (published on line), DOI: 10.1002/ird.385.

[243] Srinivasulu, A, Satyanarayana, TV, and Hema Kumar, HV. 2005. Subsurface drainage in a pilot area in Nagarjuna Sagar right canal command, India. *Irrigation and Drainage Systems*, **19**, 61-70.

[244] Srinivasulu, A, Sujani Rao, C, Lakshmi, GV, Satyanarayana, TV, and Boonstra, J. 2004. Model studies on salt and water balances at Konanki pilot area, Andhra Pradesh, India. *Irrigation and Drainage Systems*, **18**, 1-17.

[245] Stuyt, L, Dierickx, W, and Beltrán, JM. 2000. Materials for subsurface land drainage systems. FAO Irrigation and Drainage Paper 60, Food and Agriculture Organization of the United Nations.

[246] Stuyt, LCPM, and Dierickx, W. 2006. Design and performance of materials for subsurface drainage systems in agriculture. *Agricultural Water Management*, **86** (1-2), 50 - 59. DOI: 10.1016/j.agwat.2006.06.004.

[247] Stuyt, LCPM, and Willardson, LS. 1999. Drain envelopes. In: RW Skaggs and J van Schilfgaarde (Eds). Agricultural drainage, American Society of Agronomy, 927-962.

[248] Svendsen, M, and Meinzen-Dick, R. 1997. Irrigation management institutions in transition: a look back, a look forward. *Irrigation and Drainage*, **11**, 139-156.

[249] Tanji, KK, and Kielen, NC. 2002. Agricultural drainage water management in arid and semi-arid areas. Food and Agriculture Organization of the United Nations, Rome.

[250] Tardieu, H. 2004. Irrigation and drainage services - some principles and issues towards sustainability. ICID Position Paper, International Commission on Irrigation and Drainage.

[251] Team, CD. 1984. Subsurface drainage design analysis. Mardan Salinity Control and Reclamation Project, Water and Power Development Authority, Lahore, Pakistan.

[252] Thomas, S. 2002. *What is participatory learning and action: an introduction.* Centre for International Development and Training., Wolverhampton.

[253] Trostle, R. 2008. Global agricultural supply and demand: factors contributing to the recent increase in food commodity prices. United States Department of Agriculture, Economic Research Service, Washington

[254] Tyagi, NG. 2002. Drainage water reuse and disposal in Northwest India. In: KK Tanji and NC Kielen (Eds). Agricultural drainage water management in arid and semi-arid areas, Irrigation and Drainage Paper no. 61, Food and Agriculture Organization of the United Nations, Rome, 114-116.

[255] Uijterlinde, R. 2007. Decentralised water management in the Netherlands: Waterschappen. *Land and Water Management in Europe*, **ERWG Letter 17**, 2-3.

[256] UNESCO-IHE. 2007. Draft conclusions and recommendations. Water for a changing world: enhance local knowledge and capacity, Delft, The Netherlands.

[257] United Nations. 2008. UN millennium development goals. Web service Section, Department of Public Information, United Nations, http://www.un.org/millenniumgoals/index.html, New York.

[258] United Nations Department of Economic and Social Affairs Population Division. 2007. *World population prospects: the 2006 revision, vol. I, comprehensive tables.* United Nations New York.

[259] United States Bureau of Reclamation. 1978. *Drainage manual.* Department of the Interior.

[260] Van Achthoven, T, Lohan, HS, and Bradley, WP. 2000. The reclamation of waterlogged and saline lands with subsurface drainage: an overview of the Haryana operational pilot project. 8th International Drainage Workshop, New Delhi, India, 515-528.

[261] Van de Ven, GP. 2004. Man-made lowlands: history of water management and land reclamation in the Netherlands. GP Van de Ven, ed., International Commission on Irrigation and Drainage and Royal Institute of Engineers, Utrecht: Matrijs. , 432.

[262] Van den Berg, J, Geerling-Eiff, F, and Steingröver, E. 2008. Community of practice: actieonderzoek voor duurzame gebiedsontwikkeling (in Dutch). Wageningen University and Research Centre, Wageningen.

[263] Van den Berg, S. 2004. Verdeeld land. De geschiedenis van de ruilverkaveling In Nederland vanuit een lokaal perspectief, 1890-1985 (in Dutch). PhD Thesis, Historia Agriculturae 35, Nederlands Agronomisch Historisch Instituut, Wageningen University, Groningen and Wageningen.

[264] Van der Schans, ML, and Lempérière, P. 2006. Manual - participatory rapid diagnosis and action planning for irrigated agricultural systems. APPIA Project, IWMI Sub-regional Office for the Nile Basin and Eastern Africa, International Programme for Technology and Research in Irrigation and Drainage, International Water Management Institute and Food and Agriculture Organization of the UN, Rome.

[265] Van der Zel, HJ, and Amer, MH. 1983. The Egyptian-Dutch advisory panel on land drainage: its activities and impact. ILRI Annual Report 1982, International Institute for Land Reclamation and Improvement, Wageningen, 9-22.

[266] Van Hofwegen, P. 2004. Capacity-building for water and irrigation sector management with application in Indonesia. Capacity Development in Irrigation and Drainage Issues, Challenges and the Way Ahead. FAO Water Report, 26, Food and Agriculture Organization of the United Nations, Rome.

[267] Van Hoorn, JW, and Van Alphen, JG. 2006. Salinity control. In: HP Ritzema (Ed). Drainage Principles and Applications, 3rd edition. ILRI Publication 16, Alterra-ILRI, Wageningen, 533-600.

[268] Van Schaik, H, Kabat, P, and Connor, R. 2003. Conclusions and recommendations. In: B Appleton (Ed). Climate changes the water rules: how water managers can cope with today's climate variability and tomorrow's climate change, Dialogue on Water and Climate, Delft, 94-97.

[269] Van Schilfgaarde, J. 1957. Approximate solutions to drainage flow problems. In: JN Luthin (Ed). Drainage of agricultural lands, American Society of Agronomy, 79-112.

[270] Van Schilfgaarde, J. 1979. Progress and problems in drainage design. In: J Wesseling (Ed). International Drainage Workshop, 633-644.

[271] Van Walsum, P, Helming, J, Stuyt, L, Schouwenberg, E, and Groenendijk, P. 2007. Spatial planning for lowland-stream basins using a bioeconomic model. *Environmental Modelling & Software*, **23**, 569-578 DOI:10.1016/j.envsoft.2007.08.006.

[272] Van Zeijts, TEJ, and Naarding, WH. 1990. Possibilities and limitations of trenchless pipe drain installation in irrigated areas. Installation of pipe drains, Government Service for Land and Water Use, Utrecht.

[273] Van Zeijts, TEJ, and Zijlstra, G. 1999. Quality assurance and control. In: RW Skaggs and J Van Schilfgaarde (Eds). Agricultural drainage, American Society of Agronomy, 1005-1022.

[274] Vandersypen, K, Keita, CTA, Lodon, B, Raes, D, and Jamin, JY. 2007. Didactic tools for supporting participatory water management in collective irrigation schemes. *Irrigation and Drainage Systems*, DOI: 10.1007/s10795-007-9042-2.

[275] Ven, GA. 1983. Design of subsurface drainage systems in Egypt. Report 83-XII-06. Advisory Panel for Land Drainage in Egypt, Drainage Research Institute, Cairo.

[276] Verhallen, A, Warner, J, and Santbergen, L. 2007. Towards evaluating MSPs for integrated catchment management. In: J Warner (Ed). Multi-stakeholder platforms for integrated water management, Ashgate, Hamphire, 259-271.

[277] Vietnam Institute for Water Resources. 2006. A participatory approach for the formulation and design of future drainage projects - Implementation manual (draft). Vietnam Institute for Water Resources, Hanoi.

[278] Vietnam Institute for Water Resources. 2006. Research on the effectiveness of drainage – draft final report. Vietnam Institute for Water Resources, Hanoi.

[279] Vietnam Institute for Water Resources Research. 2003. Final report on benefit monitoring and evaluation system 1999-2002. Red River Delta Water Resources Sector Project (Loan no. 1344 – VIE(SF)). Vietnam Institute for Water Resources Research, Hanoi.

[280] Vlotman, WF, Vlotman, WF, Shafiq-ur-Rehman, and Haider, I. 1993. Granular envelope research in Pakistan. *Irrigation and Drainage Systems*, **6**, 325-343.

[281] Vlotman, WF, Willardson, LS, and Dierickx, W. 2001. *Envelope design for subsurface drains*. International Institute for Land Reclamation and Improvement, Wageningen.

[282] Vlotman, WF, Wong, T, and Schultz, B. 2007. Integration of drainage, water quality and flood management in rural, urban and lowland areas. *Irrigation and Drainage*, **56**, S161-S177, DOI: 10.1002/ird.369.

[283] von Braun, J. 2008. Rising food prices - What should be done? IFPRI Policy Brief International Food Policy Research Institute, Washington.

[284] Wahba, MAS, Christen, EW, and Amer, MH. 2005. Irrigation water saving by management of existing subsurface drainage in Egypt. *Irrigation and Drainage*, **54**, 205-215.

[285] Warner, J. 2007. The beauty of the beast: multi-stakeholder participation for integrated catchment management. In: J Warner (Ed). Multi-stakeholder platforms for integrated water management, Ashgate, Hampshire, 1-20.

[286] Water Resources Consulting Services. 2000. Final report on management study on land use and water management. Water Resources Consulting Services, Asia Development Bank, TA no. 2871-VIE, Hanoi.

[287] Water Resources Planning Organisation and International Waterlogging and Salinity Research Institute. 2005. Drainage master plan of Pakistan. Water Resources Planning Organisation and International Waterlogging and Salinity Research Institute, Water and Power Development Authority, Lahore.

[288] Watson, N. 2007. Collaborative capital: a key to successful practice of integrated water resources management. In: J Warner (Ed). Multi-stakeholder platforms for integrated water management, Ashgate, Hampshire, 31-48.

[289] Weissink, A. 2007. Polder aan de Nijl - Nederlandse Waterschappen in Egypt (in Dutch). NRC Handelsblad, Zaterdags Bijvoegsel, 41.

[290] Wesseling, J. 1983. Subsurface flow into drains. Drainage principles and applications, International Institute for Land Reclamation and Improvement, Wageningen, 1-56.

[291] Wolters, W. 2000. Research on technical drainage issues: lessons learned from the IWASRI/NRAP project 1988-2000. Publication 226, International Waterlogging and Salinity Research Institute, Lahore.

[292] Wolters, W, Bhutta, MH, and Plusquellec, H. 2006. Applied water research: value for money. GRID, IPTRID Network Magazine, HR Wallingford, UK, 18-20.

[293] Wolters, W, Habib, Z, and Bhutta, MH. 1997. Forget about crop-demand based canal irrigation in the plains of Pakistan. Proceedings Seminar on-farm salinity, drainage, and reclamation. IWASRI Publication 179., Lahore.

[294] Wolters, W, Ittfaq, M, and Bhutta, MH. 1996. Drainage discharge and quality for drainage options in Pakistan. Workshop on Managing Environmental Changes due to Irrigation and Drainage, 16th Congress on Irrigation and Drainage, Cairo, 44-55.

[295] Wolters, W, Ritzema, HP, and Maaskant, M. 1986. Polders in Egypt. *Land + Water International* **58**, 35-40.

[296] World Bank. 2004. Drainage for gain - integrated solutions to drainage in land and water management. Agriculture and Rural Development Department, The World Bank, Washington.

[297] World Bank. 2004. Egypt: improving agricultural production through better drainage. Agriculture Investment Sourcebook, Module 08, Investments in Irrigation and Drainage, World Bank, Washington.

[298] World Bank. 2006. Re-engaging in agricultural water management: challenges and options. Direction in development, World Bank, Washington.

[299] World Bank. 2007. *Enhancing agricultural innovation - how to go beyond the strengthening of research systems.* The World Bank, Washington.

[300] World Bank. 2008. World development report 2008 - agriculture for development. The International Bank for Reconstruction and Development / The World Bank, Washington.

[301] WRPO, and IWASRI. 2005. Drainage master plan of Pakistan. Water Resources Planning Organisation and International Waterlogging and Salinity Research Institute, Water and Power Development Authority, Lahore.

[302] Zwarteveen, M. 2006. Effective gender mainstreaming in water management for sustainable livelihoods: from guidelines to practice. Both Ends Working Paper Series, Amsterdam.

Abbreviations and acronyms

AAPk	Action Aid Pakistan
ANGRAU	Acharya N.G. Ranga Agricultural University
APP	Advisory Panel Project
AWB	Area Water Boards
Berseem	Egyptian clover
BOD	Bio-oxygen demand
CADA	Command Area Development Authority
CCAD	Chashma Command Area Development project
CSSRI	Central Soil Salinity Research Institute
CWR	Crop water requirement
DARD	Department of Agriculture and Rural Development
DEMP	Drainage Executive Management Project
DO	Dissolved oxygen
DRI	Drainage Research Institute
DTAP	Drainage Technology and Pilot Areas
DTC	Drainage Training Centre
EC	Electrical conductivity
EKTDP	East Khairpur Tile Drainage Project
EPADP	Egyptian Public Authority for Drainage Projects
FDP	Fourth Drainage Project
FESS	Fordwah Eastern Sadiqia South Project
FO	Farmer Organizations
Ha	Hectare (10,000 m^2)
GDP	Gross domestic product
HOPP	Haryana Operational Pilot Project
ICAR	Indian Council of Agricultural Research
IDMC	Irrigation and Drainage Management Committees
IDNP	Indo-Dutch Network Operational Research Project on Drainage and Water Management for Control of Salinity and Waterlogging in Canal Commands
ILRI	International Institute for Land Reclamation and Improvement
IPTRID	International Programme for Technology and Research in Irrigation and Drainage
IRR	Economic internal rates of return
IWASRI	International Waterlogging and Salinity Research Institute
IWRM	Integrated Water Resource Management
Kharif	Monsoon season in South Asia (from July to October)
KLURE	Kolleru Lake and Upputeru River Ecosystem research project
KOMO	*Stichting voor Onderzoek, Beoordeling en Keuring van Materialen en Constructies* (in Dutch)
LBOD	Left Bank Outfall Drain
MALR	Ministry of Agriculture and Land Reclamation
MARDAN	Mardan Salinity Control and Reclamation Project
MDG	Millennium Development Goals
MSL	Mean sea level
MSP	Multi-stakeholder platform

MWRI	Ministry of Water Resources and Irrigation
Nala	Natural stream or drainage channel
NGO	Non-governmental organization
NRAP	Netherlands Research Assistance Project
NWRC	National Water Research Centre
O&M	Operation and maintenance
ORU	Operational Research Unit of EPADP
PDM	Participatory Drainage Management
PEC	Public Excavation Companies
PFD	Planning and Follow-up Department of EPADP
PID	Provincial Irrigation Department
PIDA	Provincial Irrigation and Drainage Authority
PLA	Participatory learning and action
RAJAD	Rajasthan Agricultural Drainage Research Project
Rabi	Post-monsoon or winter season in South Asia (from October to March)
RIGW	Research Institute for Ground Water
RIJP	Rijksdienst voor de IJsselmeerpolders / *IJsselmeerpolders* Development Authority
RRWRSP	Red River Delta Water Resources Sector Project
SAR	Sodium absorption ratio
SCARP	Salinity Control and Reclamation Projects
SDC	Sub-project Drainage Committees
SRKR	Sagi Ramakrishnam Raju Engineering College
SRRBSP	Second Red River Basin Sector Project
SS	Soil surface
SWERI	Soils, Water and Environment Research Institute
SWOT	Strength-Weakness-Opportunities-Threat analysis
TO	Total solids
USBR	United Stated Bureau of Reclamation
WAPDA	Water and Power Development Authority
WG-CBTE	Working Group on Capacity Building, Training and Education
Zaid	Summer season in South Asia (March to June)

List of symbols

Symbol	Description	Units
A	Cross-sectional area, drained area	m^2, km^2
B/C	Benefit / Cost ratio	-
C	Salt concentration	$dS\ m^{-1}$
D	Drainage through subsurface drainage system	mm, $mm\ d^{-1}$
d	Diameter	mm, m
d, D	Depth, equivalent depth, thickness, height	M
E	Evaporation	mm, $mm\ d^{-1}$
EC	Electrical conductivity at 25 ^{0}C	$dS\ m^{-1}$
F	Frequency	-
G	Capillary rise	mm, $mm\ d^{-1}$
h, H	Water depth	m
h, H	(Energy) head or head loss	m
I	Irrigation	mm, $mm\ d^{-1}$
K_m	Manning's roughness coefficient	-
L	Length, spacing, width	m
LF	Leaching fraction	-
N	Natural drainage	mm, $mm\ d^{-1}$
O_{90}	Pore size of envelope retaining 90% of soil fraction	m
P	Precipitation	mm, $mm\ d^{-1}$
q, Q	Discharge, flow rate, runoff rate	$m^3\ d^{-1}$, $m^2\ d^{-1}$, $m\ d^{-1}$
q	Drainage coefficient, drainable surplus	$mm\ d^{-1}$
q/h	Drainage intensity ratio	d^{-1}
R	Percolation	mm, $mm\ d^{-1}$
SAR	Sodium Adsorption Ration	$Meq^{0.5}/l^{0.5}$
T, t	Time, period	Yr, d, s
TS	Total salts	$t\ ha^{-1}$
W	Soil moisture	mm
W	Water storage	mm, $mm\ d^{-1}$
W	Watt	$J\ s^{-1}$, $N\ m\ s^{-1}$
Y	Yield	$t\ ha^{-1}$
Z	Amount of salt	$mm\ dS\ m^{-1}$, $t\ ha^{-1}$

Samenvatting

De wereldbevolking zal naar verwachting groeien van 6500 miljoen nu, naar 9100 miljoen in 2050. Om voor deze populatie voldoende voedsel te produceren zal de productie moeten verdubbelen. Deze verdubbeling moet hoofdzakelijk komen uit opbrengstverhogingen, omdat er weinig mogelijkheden zijn overgebleven het huidige landbouwareaal te vergroten. Het aandeel van de geïrrigeerde landbouw in de wereldvoedselproductie zal naar verwachting toenemen van 35-45% naar meer dan 50%. Deze vorm van landbouw heeft echter te kampen met verzilting en/of wateroverlast, met als gevolg dat er jaarlijks ongeveer een half miljoen hectare landbouwgrond uit productie wordt genomen. Om problemen van verzilting en wateroverlast aan te pakken, zouden op ongeveer 60 miljoen hectare drainagesystemen moeten worden aangelegd, worden vervangen of gemoderniseerd. Van deze systemen zal de helft bestaan uit buizendrainage. Met een gemiddelde kostprijs van € 1250 per hectare, betekent dit een wereldwijde investering van om en nabij 19 miljard Euro ofwel de komende 40 jaar 475 miljoen Euro per jaar.

Opzet van de studie
In deze studie wordt voor de geïrrigeerde landbouw in de aride - en semi-aride regio's de rol van de drainage geanalyseerd en worden aanbevelingen geformuleerd om de bestaande systemen te verbeteren. Gebaseerd op de kennis die ik de afgelopen 28 jaar heb opgedaan in Egypte, India en Pakistan, aangevuld met ervaringen uit andere landen waar ik gewerkt heb, beschrijf ik de geleidelijke verandering van een monodisciplinaire naar een multidisciplinaire aanpak. Ik heb de drainage niet afzonderlijk bekeken maar in haar relatie tot integraal waterbeheer, waarbij behalve de technische, ook de sociaaleconomische en organisatorische aspecten in beschouwing worden genomen. Dit proefschrift is een synthese van een aantal casestudies die afzonderlijk zijn gepubliceerd in internationale wetenschappelijke tijdschriften.

Een verbeterd drainagesysteem voor rijstgebieden
De eerste casestudy beschrijft de ontwikkeling van een verbeterd drainagesysteem voor landbouwgebieden in Egypte, waar de verbouw van rijst wordt afgewisseld met andere gewassen (Hoofdstuk 2.7). Het onderzoek richtte zich op de vraag hoe percolatie van irrigatiewater in rijstvelden kan worden verminderd zonder de drainage van belendende percelen te belemmeren. Allereerst werd het concept voor het verbeterde drainagesysteem getest in drie proefvelden. De uitkomsten toonden aan dat de introductie van het verbeterde drainagesysteem een waterbesparing van ongeveer 30% opleverde, zonder negatieve gevolgen voor de gewasopbrengsten en het zoutgehalte in de bodem. Vervolgens werd de haalbaarheid getest in de praktijk, waarbij de boeren zelf bepaalden welke gewassen zij verbouwden en hoe zij hun irrigatie en drainage regelden. Dit vervolgonderzoek toonde aan dat (i) de aanlegkosten ongeveer gelijk waren aan die van de aanleg van het conventionele drainagesysteem, (ii) de boeren zelfs nog meer irrigatiewater (en dus pompkosten) bespaarden dan verwacht, (iii) er minder schade voor andere gewassen optrad omdat het drainagesysteem niet meer (illegaal) werd geblokkeerd, (iv) dit resulteerde in lagere beheer- en onderhoudskosten.

Verifiëren van de ontwerpnormen voor drainagesystemen
De tweede casestudy beschrijft de resultaten van een onderzoekprogramma dat werd uitgevoerd om de ontwerpcriteria voor drainagesystemen te verifiëren in Mashtul, een proefveld van 110 ha in het zuidoostelijke deel van de Nijldelta in Egypte (Hoofdstuk 2.9). Dit onderzoek toonde aan dat na de aanleg van het drainagesysteem de gewasopbrengsten significant toenamen. De gemiddelde toename was 10% voor rijst, 48% voor *berseem* (Egyptische klaver), 75% voor maïs en meer dan 130% voor de tarwe. Deze opbrengstverbeteringen konden worden toegeschreven aan (i) de verlaging van het zoutgehalte in de bodem, (ii) een verbeterde lucht- en waterhuishouding in de wortelzone en (iii) een verbeterde landbouwtechnische aanpak. Verder kon geconcludeerd worden dat de bestaande drainagesystemen waren overgedimensioneerd. Vervolgonderzoek toonde aan dat een betere afstemming tussen irrigatie en drainage resulteert in zowel een lagere irrigatiebehoefte als in een verminderde drainageafvoer zonder dat dit negatieve effecten heeft op de gewasopbrengst en het zoutgehalte in de bodem.

Ontwikkelen van drainagestrategieën
De derde casestudy beschrijft het ontwikkelen van drainagestrategieën voor de geïrrigeerde landbouw in vijf verschillende agro-klimatologische regio's in India (Hoofdstuk 3.6). Uit het onderzoek bleek dat onder de specifieke bodemkundige, agronomische, sociaaleconomische en klimatologische omstandigheden, een samengesteld drainagesysteem (bestaande uit open sloten of ondergrondse buizen) een technisch en financieel interessante en sociaal acceptabele methode is om wateroverlast en verzilting te voorkomen. Na aanleg van de drainagesystemen gingen de gewasopbrengsten binnen twee seizoenen omhoog, gemiddeld 54% voor suikerriet, 64% voor katoen, 69% voor rijst en 136% voor tarwe. Deze opbrengsten konden worden behaald omdat de grondwaterstanden en de zoutgehaltes in de bodem gemiddeld 25% en 50% lager waren dan in de niet-gedraineerde gebieden. Op basis van de onderzoekresultaten werden drainagestrategieën voor de vijf agro-klimatologische regio's geformuleerd. De aanbevolen diepte, variërend tussen 0.5 en 1.5m, van de drainagesystemen is significant ondieper dan tot nu toe gebruikelijk (> 1.75m).

De rol van participatief modelleren in drainageonderzoek
De vierde casestudy was erop gericht consensus te kweken voor een integrale aanpak van het ecologisch herstel van het Kolleru-Upputeru wetland, een dichtbevolkt gebied aan de oostkust van Andhra Pradesh in India (Hoofdstuk 3.9). Het Kolleru Lake-gebied omvat niet alleen een natuurreservaat (RAMSAR site), maar is ook gedeeltelijk ingepolderd. Het water is het grootste deel van het jaar zoet en wordt dan gebruikt voor irrigatie, maar in de droge tijd dringt het zoute zeewater via de Upputeru River het gebied binnen. Ook vervuilt het meer door de lozing van grote hoeveelheden huishoudelijk, industrieel en landbouwkundige afvalwater. In de studie werd één van de belangrijkste beperkingen van het gebruik van modellen, namelijk de noodzaak voor langjarige meetreeksen, opgelost door de lokale kennis van de inwoners te combineren met de expliciete kennis van de onderzoekers. Door de modeluitkomsten te bespreken kregen de belanghebbenden een beter beeld van de complexiteit van de problematiek en werd hun ook duidelijk dat single-issue oplossingen niet werken. Zo ontstond er een beter begrip voor elkaars belangen en werd overeenstemming bereikt over de noodzaak van een integrale aanpak.

De ontwikkeling van een bedrijfstak
De vijfde casestudy illustreert hoe, over de laatste 50 jaar, de drainagepraktijk zich heeft ontwikkeld van een kleinschalige, arbeidsintensieve methode tot een grootschalige, gemechaniseerde bedrijfstak (Hoofdstuk 5.1). Deze transformatie werd mogelijk door de ontwikkeling van nieuwe installatietechnieken, machines, materialen, planning- en ontwerpmethoden en organisatievormen. De kwaliteit van de deze grootschalige, gemechaniseerde installatieprocessen werd gewaarborgd door de ontwikkeling van een integraal kwaliteitscontrolesysteem. Al deze veranderingen waren mogelijk omdat behalve aan onderzoek ook veel aandacht is besteed aan opleiding. Naast de traditionele "klassikale" onderwijsmethoden bleek een meer praktische "op locatie" training bijzonder effectief.

De toegevoegde waarde van drainageonderzoek
In de zesde casestudy wordt de rol van onderzoek in de ontwikkeling van nieuwe methoden en technieken besproken (Hoofdstuk 5.2). Er is geanalyseerd hoe onderzoek heeft bijgedragen aan het verbeteren van de bedrijfsvoering, inclusief organisatorische aanpassingen. Daarnaast is onderzocht of deze innovaties ook werkelijk werden geaccepteerd en is er een schatting gemaakt van de kostenbesparingen die de invoering van deze innovaties zouden hebben opgeleverd. De resultaten van dit onderzoek worden besproken voor de vier fases van het drainageproces, te weten: (i) voorbereiding, (ii) planning en ontwerp, (iii) aanleg en (iv) beheer en onderhoud.

Een integrale aanpak voor capacity building in drainage (en irrigatie)
Nieuwe en verbeterde technieken en methoden konden worden geïntroduceerd omdat niet alleen werd geïnvesteerd in onderzoek, maar ook in opleiding. In deze casestudy wordt een integrale aanpak voor capacity building voor drainage (en irrigatie) gepresenteerd, die is gebaseerd op ervaringen uit Nederland, Egypte, Indonesië, Pakistan, Indonesië en Maleisië (Hoofdstuk 6.1). Capacity building is een cyclisch leerproces waarin de meer expliciete of tastbare kennisaspecten, zoals basisprincipes, wetenschappelijke kennis, handboeken, enz., worden gekoppeld aan de intrinsieke of lokaal beschikbare kennis. Onderzoek, training en advisering zijn essentiële elementen in het proces van capacity building. *Onderzoek* is noodzakelijk om lokale kennis te koppelen aan expliciete kennis van elders en zo nieuwe kennis te genereren om de lokale problemen op te lossen. *Onderwijs en training* zijn nodig om deze nieuwe kennis te implementeren. Daarbij bieden onderwijs en training de mogelijkheid om de intrinsieke kennis van de cursisten expliciet te maken. Ten slotte is a*dvisering* noodzakelijk om de nieuwe opgedane kennis toe te passen. Hiermee is het cyclische leerproces doorlopen, maar kennisopbouw wordt efficiënter naarmate dit proces vaker wordt herhaald. Dit impliceert een langdurige samenwerking, waardoor het onderlinge vertrouwen kan groeien en de efficiëntie van capacity building wordt vergroot.

De rol van participatief onderzoek in de voorbereiding van drainage projecten
De achtste en laatste casestudy beschrijft een participatief onderzoek dat werd uitgevoerd in het kader van een programma om maatregelen te ontwikkelen die de drainage in twee polders in de delta van de Red River in Vietnam moesten verbeteren (Hoofdstuk 6.2). Niet alleen de boeren, maar ook de andere bewoners uit deze dichtbevolkte polders en de relevante overheids- en particuliere organisaties werden bij de studie betrokken. Om de afwatering van overtollig regenwater in deze voor de delta zo kenmerkende, complexe poldersystemen aan te pakken, bleek behalve een verbetering van de technische infrastructuur ook een goede samenwerking tussen de belanghebbenden onontbeerlijk. Het

onderzoek werd afgesloten met workshops waarin de betrokken partijen tot overeenstemming kwamen over een gefaseerde invoering van de voorgestelde verbeteringen. Hoewel dit onderzoek werd uitgevoerd in de humide tropen, kan deze participatieve planningsmethodiek ook worden gebruikt in aride- en semi-aride regio's, aangezien de drainageproblematiek in deze gebieden een overeenkomstige technische en organisatorische complexiteit vertoont.

Synthese: de praktijk van drainage in de geïrrigeerde landbouw

In de synthese wordt uiteengezet welke maatregelen op het gebied van drainage nodig zijn om de voedselproductie in de geïrrigeerde landbouw in de aride - en semi-aride regio's te bevorderen (Hoofdstuk 7). Analyse van de bestaande praktijken in Egypte, India en Pakistan toont aan dat de bestaande drainagesystemen in technische zin voldoen. Deze systemen voorkomen hoge grondwaterstanden en verzilting en hebben daardoor een positief effect op de gewasopbrengsten en de inkomens van de boeren. Het onderzoek laat zien dat diepe drainagesystemen niet nodig zijn om verzilting tegen te gaan. Door een betere integratie van irrigatie en drainage kan de intensiteit van de drainage worden verminderd. Dit bespaart niet alleen irrigatiewater maar leidt ook tot lagere kosten en vermindert de afvoer van het (zoute) drainagewater. Het toepassen van nieuwe installatietechnieken en het gebruik van nieuwe materialen voor de buizen en omhullingmaterialen maken het mogelijk drainage op een meer efficiënte manier aan te leggen. De kosten-batenanalyses tonen aan dat deze systemen rendabel zijn. De huidige stijging van de prijzen van landbouwproducten zal de economische rentabiliteit nog verder vergroten.

Hoewel geconcludeerd kan worden dat de bestaande drainagesystemen technisch voldoen en kostendekkend zijn, blijft de ontwikkeling van drainage echter ver achter bij de ontwikkeling van de irrigatie. Dit heeft als gevolg dat grote gebieden te maken hebben met wateroverlast en verzilting. Egypte, waar de overheid de verantwoordelijkheid voor de ontwikkeling van drainage op zich nam, vormt hierop een uitzondering. Maar zelfs in Egypte blijkt het problematisch om het beheer van drainagesystemen over te dragen aan de boeren of hun belangenorganisaties. De belangrijkste oorzaak is dat de systemen zijn ontworpen en aangelegd door de overheid. Bij de totstandkoming zijn de gebruikers, hoofdzakelijk kleine boeren, niet of nauwelijks betrokken geweest. In deze 'top-down' aanpak wordt te weinig rekening gehouden met de belangen van de boeren en de lokale omstandigheden, waardoor er bij de boeren onvoldoende verantwoordelijkheidsbesef ontstaat voor het beheer en onderhoud van het systeem. Daarnaast lag bij de aanleg het accent vooral op de technische aspecten (de fysieke infrastructuur) terwijl de organisatorische aspecten (de institutionele infrastructuur) ondergeschikt bleven. Er is echter hoop, want hoewel de meeste boeren te arm zijn om de noodzakelijke investeringen in drainage op te brengen, zijn zij wel overtuigd van de voordelen en ook bereid om bij te dragen: financieel of door middel van het leveren van arbeid.

Hoe nu verder: aanbevelingen om de rol van drainage in de geïrrigeerde landbouw te versterken

Om de negatieve tendens van wateroverlast en verzilting in de geïrrigeerde landbouw te keren, heb ik een drietal aanbevelingen geformuleerd om de rol van drainage te versterken (Hoofdstuk 8). Deze aanbevelingen zijn: (i) een betere balans tussen "*top-down*" en "*bottom-up*" (ii) van standaardisatie naar een meer flexibele aanpak en (iii) focus op capacity building.

(i)   *Een betere balans tussen "top-down" en "bottom-up".* Boeren en andere belanghebbenden moeten betrokken worden bij elke fase van het drainageproces. Dit begint met een inventarisatie van hun problemen, hun voorkeuren en de bereidheid bij te dragen aan een oplossing. De 'participatory learning and action' aanpak is een effectieve en efficiënte methode om de noodzaak van drainage te bepalen, om begrip voor elkaars problemen te kweken en om draagvlak voor een integrale aanpak te creëren. Participatieve onderzoek- en modelleringtechnieken helpen de belanghebbenden de complexiteit van de problematiek beter te begrijpen en geven een beter inzicht in de effectiviteit van de verschillende oplossingsrichtingen. Deze participatieve methoden zijn ook uitermate geschikt om de samenhang te laten zien tussen de technische aspecten (die om fysieke oplossingen vragen) en de organisatorische aspecten (die om institutionele oplossingen vragen).

(ii)   *Van standaardisatie naar een meer flexibele aanpak.* In plaats van de gebruikelijke gestandaardiseerde ontwerp- en installatietechnieken wordt een meer flexibele aanpak aanbevolen zodat beter rekening gehouden kan worden met de belangen van de gebruikers en de lokale omstandigheden. Ook is het nodig om tot een verdere integratie van de irrigatie- en drainagesystemen te komen. De uitdaging hierbij is een balans te vinden tussen individuele en collectieve belangen. Enerzijds is er de gewenste intensiteit van drainage, die van veld tot veld en van boer tot boer zal verschillen; anderzijds is drainage een collectieve activiteit, waarin boeren moeten samenwerken. Eén van de technische mogelijkheden om dit te bereiken is het inbouwen van meer individuele controlemogelijkheden in het drainagesysteem. Dit geeft de boeren de mogelijkheid de waterhuishouding in hun veld te optimaliseren. Tevens geeft deze grotere flexibiliteit de boeren de gelegenheid te anticiperen en te reageren op veranderingen in het landgebruik en/of de gevolgen van veranderingen in het klimaat.

(iii)   *Focus op capacity building.* Om de participatie van de belanghebbenden te verbeteren en tot een meer flexibele aanpak te komen is capacity building essentieel. De expliciete kennis van de onderzoekers moet gekoppeld worden aan de intrinsieke of lokale kennis van de belanghebbenden. Door deze twee soorten kennis te verenigen wordt het noodzakelijke nieuwe inzicht gegenereerd die nodig is om de problemen aan te pakken. Op haar beurt moet deze kennis expliciet gemaakt worden om via onderwijs, training en voorlichting overgedragen te kunnen worden aan de belanghebbenden. In dit proces van capacity building zijn onderzoek, training en advisering (het zogenaamde OVO-drieluik) essentiële elementen.

Ik ben ervan overtuigd dat bovengenoemde aanbevelingen een belangrijke bijdrage kunnen leveren aan een meer duurzaam gebruik van bodem en water, natuurlijke hulpbronnen die steeds verder uitgeput raken. De boeren moeten zich bewust worden dat ook zij dienen te investeren in deze hulpbronnen. In samenspraak met de belanghebbenden zal de overheid een beleid moeten ontwikkelen waarin de bestrijding van wateroverlast en verzilting dezelfde prioriteit krijgt als de veroorzaker van het probleem: de irrigatie. Dit laat onverlet dat niet alleen de boeren maar ook de andere belanghebbenden bijdragen in de kosten: via betaling of door het leveren van arbeid.

Onderzoek zal nodig blijven om tegemoet te komen aan de specifieke wensen en behoeften in de zich ontwikkelende landen, elk met zijn specifieke klimatologische, fysieke en sociaaleconomische omstandigheden. Aspecten zoals klimaatverandering, veranderend landgebruik en de veranderde eisen met betrekking tot de beschikbaarheid en de kwaliteit van het water moeten hierbij worden meegenomen. Wij moeten ons blijven realiseren dat alleen de boer in staat is om zijn dagelijkse praktijk aan te passen aan deze veranderende omstandigheden. Het is de uitdaging voor onderzoekers en opleiders om de boeren in staat te stellen hun land op een duurzame manier te bewerken en ze tegelijkertijd te leren inspelen op bovengenoemde veranderingen. Alleen wanneer wij deze uitdaging oppakken zullen de investeringen in de geïrrigeerde landbouw in aride- en semi-aride gebieden hun nut bewijzen en bijdragen aan de voedselvoorziening van de almaar groeiende wereldbevolking.

Bibliography

Books and chapters in books

Abdel-Dayem, S., Ritzema, H. P., El-Atfy, H. E., and Amer, M. H., 1989. Pilot Areas and Drainage Technology. In: M. H. Amer and N. A. de Ridder, (eds). Land Drainage in Egypt, Drainage Research Institute, Cairo, 103-161.

De Glopper, R. J., and Ritzema, H. P., 1994. Land subsidence. In: H. P. Ritzema, ed. Drainage Principles and Applications, 2nd completely revised edition, Alterra-ILRI, Wageningen, 477-512.

Nijland, H. J., Croon, F. W., and Ritzema, H. P., 2005. Subsurface Drainage Practices: Guidelines for the implementation, O&M of subsurface pipe drainage systems. ILRI Publication 60, Alterra, Wageningen University and Research Centre, Wageningen.

Rieley, J.O. and S.E. Page (Eds). 2005. Wise Use of Tropical Peatlands: Focus on Southeast Asia. STRAPEAT and RESTORPEAT Project, Alterra, Wageningen University and Research Centre, The Netherlands, 231 pp. (Collaborating and contributing partner).

Ritzema, H. P. (Editor-in-Chief), 1994. Drainage Principles and Applications. Second completely revised edition. ILRI Publication 16, International Institute for Land Reclamation and Improvement, Wageningen, 1125.

Ritzema, H. P., 1994. Subsurface flow to drains. In: H. P. Ritzema, ed. Drainage Principles and Applications, 2nd completely revised edition, Alterra-ILRI, Wageningen,, 263-304.

Ritzema, H. P., and Braun, H. M. H., 1994. Environmental aspects of drainage. In: H. P. Ritzema, ed. Drainage Principles and Applications. ILRI Publication 16., International Institute for Land Reclamation and Improvement, Wageningen, 1041-1066.

Ritzema, H. P., Kselik, R. A. L., and Chanduvi, F., 1996. Drainage of Irrigated Lands. In: Irrigation water management training manual, no. 9, Food and Agriculture Organization of the United Nations, Rome, 74.

Articles published in International Journals

Abdel-Dayem, M. S., and Ritzema, H. P., 1990. Verification of Drainage Design Criteria in the Nile Delta, Egypt. *Irrigation and Drainage Systems,* **4**, 117-131.

El-Atfy, H. E., Abdel-Alim, M. Q., and Ritzema, H. P., 1991. A modified layout of the subsurface drainage system for rice areas in the Nile Delta, Egypt. *Agricultural Water Management,* **19**, 289-302.

Page, S., Hosciło, A., Wösten, J. H. M., Jauhiainen, J., Silvius, M., Rieley, J., Ritzema, H. P., Tansey, K., Graham, L., Vasander, H., and Limin, S., in press. Restoration ecology of lowland tropical peatlands in southeast Asia – current knowledge and future research directions. *Ecosystems.*

Park, S. H., Simm, J., and Ritzema, H. P., in press. Development of tidal areas, some principles and issues toward sustainability. *Irrigation and Drainage.*

Ritzema, H. P., Mutalib Mat Hassan, A., and Moens, R. P., 1998. A new approach to water management of tropical peatlands: a case study from Malaysia. *Irrigation and Drainage Systems,* **12**, 123-139.

Ritzema, H. P., Nijland, H. J., and Croon, F. W., 2006. Subsurface Drainage Practices: From Manual Installation to Large-Scale Implementation. *Agricultural Water Management*, **86**, 60-71.

Ritzema, H. P., Ramakrishna Rajub, Ch. Sreenivas, Kselik, R. A. L., and Froebrich, J., 2008. Participatory modelling to increase stakeholder participation in the restoration of the Kolleru – Upputeru wetland ecosystem in India. *Environmental Modelling & Software*, (submitted 08-08-2008).

Ritzema, H. P., Satyanarayana, T. V., Raman, S., and Boonstra, J., 2008. Subsurface drainage to combat waterlogging and salinity in irrigated lands in India: lessons learned in farmers' fields. *Agricultural Water Management*, **95**, 179 - 189, DOI: 10.106/j.agwat.2007.09.012.

Ritzema, H. P., Thinh, L. D., Anh, L. Q., Hanh, D. N., Chien, N. V., Lan, T. N., Kselik, R. A. L., and Kim, B. T., 2008. Participatory research on the effectiveness of drainage in the Red River Delta, Vietnam. *Irrigation and Drainage Systems*, **22**, 19-34, DOI 10.1007/s10795-007-9028-0.

Ritzema, H. P., Wolters, W., Bhutta, M. N., Gupta, S. K., and Abdel-Dayem, S., 2007. The Added Value of Research on Drainage in Irrigated Agriculture. *Irrigation and Drainage*, **56**, S205-S215, DOI: 10.1002/ird.337.

Ritzema, H. P., Wolters, W., and Terwisscha van Scheltinga, C. T. H. M., 2008. Lessons learned with an integrated approach for capacity development in agricultural land drainage. *Irrigation and Drainage*, 57, 354-365, DOI: 10.1002/ird.431.

Wolters, W., Ritzema, H. P., and Maaskant, M., 1986. Polders in Egypt. *Land + Water International*, **58**, 35-40.

Wösten, J. H. M., and Ritzema, H. P., 2001. Land and Water Management Options for Peatland Development in Sarawak. *International Peat Journal*, **11**, 59-66.

Articles published in conference proceedings, etc.

Abdalla, M. A., Abdel-Dayem, S., and Ritzema, H. P., 1990. Subsurface drainage rates and salt leaching for typical field crops in Egypt. Symposium on Land Drainage for Salinity Control in Arid and Semi-Arid Regions, Cairo, 383-392.

Abdel-Dayem, S., and Ritzema, H. P., 1987. Covered drainage systems in areas with rice in the crop rotation. International Winter Meeting of the American Society of Agricultural Engineers, Chicago, 7.

Abdel-Dayem, S., and Ritzema, H. P., 1987. Subsurface Drainage Rates and Salt Leaching in Irrigated Fields. 3rd International Workshop on Land Drainage, Columbus, Ohio, F:1-8.

Abdel-Dayem, S., and Ritzema, H. P., 1990. Verification of the drainage design criteria in Egypt. Symposium on Land Drainage for Salinity Control in Arid and Semi-Arid Regions, Cairo, 300-312.

Bos, M. G., Ritzema, H. P., Chong, T., and Liong, T. Y., 2003. Coastal Peat Swamp Development in Sarawak – Research Needs. Integrated Peatland Management for sustainable development - A Compilation of Seminar Papers, Damai, Sarawak, Malaysia, 287-296.

Diemont, H., Ritzema, H. P., Schrijver, R., Verhagen, J., Verwer, C., and Wösten, J., 2008. Spatial policy and the issue of carbon emission in peat lands International Symposium and Workshop on Tropical Peatlands 'Peatland development: wise use and impact management', Kuching, Sarawak.

Diemont, W. H., Ferwerda, W., Joosten, H., Minaeva, T., Rieley, J., Ritzema, H. P., and Silvius, M. J., 2004. The global peatland initiative as a partnership. Wise use of peatlands, 12th Int. Peat Congress, 6-11 June 2004, Tampere, Finland, 533-537.

Diemont, W. H., Hillegers, P. J. M., Joosten, H., Kramer, K., Ritzema, H. P., Rieley, J., and Wösten, J. H. M., 2002. Fire and Peat Forests, What are the Solutions? Workshop on Prevention & Control of Fire in Peatlands, Kuala Lumpur, 24.

El-Atfy, H. E., Abdel Alim, M. Q., and Ritzema, H. P., 1990. Experiences with a drainage system for rice areas in Egypt. Symposium on Land Drainage for Salinity Control in Arid and Semi-Arid Regions, Cairo, 129-141.

El-Atfy, H. E., Wahid El-Din, O., El-Gammal, H., and Ritzema, H. P., 1990. Hydraulic performance of Subsurface Collector Drains in Egypt. Symposium on Land Drainage for Salinity Control in Arid and Semi-Arid Regions, Cairo, 393-404.

Kselik, R. A. L., Smilde, K. W., Ritzema, H. P., Subagyono, K., Saragih, S., Damanik, M., and Suwardjo, H., 1993. Integrated research on water management, soil fertility and cropping systems on acid sulphate soils in South Kalimantan, Indonesia. Selected papers of the Ho Chi Minh City Symposium on acid sulphate soils, 177-194.

Kumbhare, P. S., and Ritzema, H. P., 2000. Need, Selection and Design of Synthetic Envelopes for Subsurface Drainage in Clay and Sandy Loam Soils of India. 8th ICID International Drainage Workshop, New Delhi, India, VI163-VI175.

Oosterbaan, R. J., and Ritzema, H. P., 1992. Hooghoudt's drainage equation, adjusted for entrance resistance and sloping land. 5th International Drainage Workshop, Lahore, 18-28.

Ritzema, H. P., 1985. The automatization of data collection and processing in the field of drainage research at the 'Rijksdienst voor de IJsselmeerpolders'. Rijksdienst voor de IJsselmeerpolders, Lelystad.

Ritzema, H. P., 2003. Subsurface Drainage: The forgotten factor in agricultural water management. Seminar on Drainage Need for Reclamation of Waterlogged and Saline Soils of Irrigated Lands, Gujarat Agricultural University, Navsari, Gujarat, India.

Ritzema, H. P., 2004. Water management: the key for sustainable management of tropical peatlands. Grid IPTRID Network Magazine, 10.

Ritzema, H. P., 2007. Ecosystem management in tidal areas: Ecological restoration on lakes in the Netherlands – concepts and dilemmas. INWEPF and ICID WG-STDA Joint Workshop Sustainable Paddy Farming in Asian Monsoon Region. International Network for Water and Ecosystem in Paddy Fields, Ansan, Korea, 43-58.

Ritzema, H. P., 2007. Performance Assessment of Subsurface Drainage Systems - Case Studies from Egypt and Pakistan. Alterra, Wageningen.

Ritzema, H.P. 2008. Coastal lowland development: coping with climate change: examples from the Netherlands. International workshop on sustainable paddy farming and wetlands, Korean INWERF Committee and ICID WG-SDTA, Changwon City, 29-30 Oct 2008, 1-24.

Ritzema, H. P., 2008. Coastal development in peatlands: a challenge or a curse – are experiences from the Netherlands useful in the tropics? International Symposium and Workshop on Tropical Peatlands 'Peatland development: wise use and impact management', Universiti Malaysia Sarawak, Kuching.

Ritzema, H. P., 2008. The role of horizontal subsurface drainage in irrigated agriculture in the semi-arid and arid regions. Proceedings of the 10th International Drainage Workshop of the ICID Working Group of Drainage, Helsinki/Tallinn, 6-11 July, 57-67.

Ritzema, H. P., in press. The role of drainage in the wise use of tropical peatlands. Carbon-climate-human interaction on tropical peatland. Proceedings of The International Symposium and Workshop on Tropical Peatland, Yogyakarta, 27-29 August 2007.

Ritzema, H. P., and Alim, M. A., 1985. Testing a rod rotator for cleaning subsurface drainage systems. Drainage Research Institute, Cairo.

Ritzema, H. P., Elbers, H. A. J., and Jallow, B. G., 1995. Rice Cultivation in Tidal Areas in the Gambia. In: ILRI Annual Report 1995, International Institute for Land Reclamation and Improvement, Wageningen, 8-23.

Ritzema, H. P., Grobbe, T., Chong, T. K. F., and Wösten, J. H. M., 2003. Decision Support System for Peatland Management in the Humid Tropics. 9th International Drainage Workshop, 10-13 September 2003, Utrecht.

Ritzema, H. P., and Jansen, H., 2008. Assessing the water balance of tropical peatlands by using the inverse groundwater modelling approach. After Wise Use – The Future of Peatlands, Proceedings of the 13th International Peat Congress, Tullamore, Ireland, 250-253.

Ritzema, H. P., and Jaya, A., 2005. Water management for sustainable wise use of tropical peatlands. Restoration and Wise Use of Tropical Peatland: problems of biodiversity, fire, poverty and water management, International Symposium on Tropical Peatland, CIMTROP, University of Palangka Raya, Palangka Raya, Indonesia, 41-52.

Ritzema, H. P., Kselik, R. A. L., Chong, T. F. K., and Liong, T. Y., 2003. A New Water Management Approach for Agricultural Development of the Lowland Peat Swamps of Sarawak. Integrated Peatland Management for sustainable development - A Compilation of Seminar Papers, Damai, Sarawak, Malaysia, 243-253.

Ritzema, H. P., Kselik, R. A. L., and Subagyono, K., 1993. Water-management strategies to ameliorate and use acid sulphate soils in the humid tropics. 15th Congress, The Hague, 1219-1235.

Ritzema, H. P., Murtedza, M., Page, S., Limin, S., and Wösten, J. H. M., 2006. Capacity Building for Sustainable Management of Peatlands in the Humid Tropics: From Research to Application. Workshop on Monitoring and Evaluation of Capacity Development Strategies in Agricultural Water Management, Kuala Lumpur, Malaysia, 73-85.

Ritzema, H. P., Murtedza, M., Suwido, L., and Page, S., 2004. New educational tools for sustainable management of peatlands in the humid tropics: The peatwise project. Wise use of peatlands, 12th Int. Peat Congress, 6-11 June 2004, Tampere, Finland, 1331-1335.

Ritzema, H. P., and Mutalib Mat Hassan, A., 1997. Water Management of Peat Lands in the Humid Tropics: a case study in Malaysia. 7th International Drainage Workshop, 17-21 November 1997, Penang, Malaysia, M7-1.

Ritzema, H. P., Raman, S., Satyanarayana, T. V., and Boonstra, J., 2003. Drainage protects irrigation investments. Alterra, Wageningen, The Netherlands.

Ritzema, H. P., Raman, S., Satyanarayana, T. V., and Boonstra, J., 2003. Drainage protects irrigation investments. Alterra-ILRI, Wageningen.

Ritzema, H. P., and Soppe, R. W. O., 2004. Drainage: A blessing or a curse. Symposium 'Water for a secure future', Changchun, China.

Ritzema, H. P., Suwido L., Kusin, K., and Jauhiainen, J., 2008. Canal blocking strategies to restore hydrology in degraded tropical peatlands in the former Mega Rice Project in Central Kalimantan, Indonesia. International Symposium and Workshop on Tropical Peatlands 'Peatland development: wise use and impact management', Kuching, Sarawak.

Ritzema, H. P., and Tuong, T. P., 1994. Water-management strategies as a tool for the sustainable use of acid sulphate soils. . Regional Workshop on the sustainable Use of Coastal Land in South-east Asia, Asian Institute of Technology, Bangkok.

Ritzema, H. P., Veltman, D., and Wösten, J. H. M., 2004. A system to support decision making for peatland management in the humid tropics. Wise use of peatlands, 12th Int. Peat Congress, 6-11 June 2004, Tampere, Finland, 720-725.

Ritzema, H. P., and Wolters, W., 2002. New frontiers in capacity building in drainage. Capacity building for drainage in North Africa, Cairo, Egypt, 10-14 March 2001, 29-38.

Ritzema, H. P., and Wösten, J. H. M., 2002. Hydrology of Borneo's Peat Swamps. . STRAPEAT Partners Workshop, STRAPEAT Project, Palangka Raya, Kalimantan, Indonesia and Sibu, Sarawak, Malaysia.

Ritzema, H. P., and Wösten, J. H. M., 2002. Water Management: The Key for the Agricultural Development of the Lowland Peat Swamps of Borneo. . International Workshop on Sustainable Development of Tidal Areas, Montreal, 101-113.

Roest, K., Ritzema, H. P., and Wolters, W., 2006. Drainage: the forgotten factor. Symposium on Irrigation Modernization: Constraints and Solutions, International Programme for Technology and Research in Irrigation and Drainage (IPTRID) of the Food and Agriculture Organization (FAO), Damascus, Syria.

Simm, J., Ritzema, H. P., and H., Park, S., 2003. Outline for ICID Handbook on Sustainable Development of Tidal Areas: Outline. Wise Use and Environmental Conservation of the Tidal Areas. Proceedings 2nd Int. Workshop on Sustainable Development of Tidal Areas Montpellier, 131-136.

Suwido H. Limin, Rieley, J. O., Ritzema, H. P., and Vasander, H., 2008. Some requirements for restoration of peatland in the former Mega Rice Project in Central Kalimantan, Indonesia: blocking channels, increasing livelihoods and controlling fires. After Wise Use – The Future of Peatlands, Proceedings of the 13th International Peat Congress, Tullamore, Ireland, 222-225.

Tan, A. K. C., and Ritzema, H. P., 2003. Sustainable Development in Peat land of Sarawak – Water Management Approach. International Conference on Hydrology and Water Resources in Asia Pacific Region, Kyoto, Japan.

Verhagen, A., Diemont, W. H., Limin, S., Rieley, J., Setiadi, B., Silvius, M. J., Ritzema, H. P., and Wösten, J. H. M., 2004. Financial mechanisms for wise use of peatlands in Borneo. Wise use of peatlands, 12th Int. Peat Congress, 6-11 June 2004, Tampere, Finland, 762-767.

Wösten, J. H. M., and Ritzema, H. P., 2001. Land and Water Management Options for Peatland Development in Sarawak. International Peat Journal, 11, 59-66.

Wösten, J. H. M., and Ritzema, H. P., 2007. Subsidence and water management of tropical peatlands. Peatlands International, 2, 38-39.

Wösten, J. H. M., and Ritzema, H. P., 2002. Challenges in land and water management for peatland development in Sarawak. International Symposium on Tropical Peatland, 22-23 August 2001, Vol. Peatlands for People: Natural Resource Functions and Sustainable Management, Jakarta, 51-55.

Wösten, J. H. M., and Ritzema, H. P., 2005. Tropische venen, een onbekend maar waardevol ecosystem (in Dutch). Alterra, Wageningen University and Research Centre, Wageningen.

Wösten, J. H. M., and Ritzema, H. P., 2005. Water management for multiple wise use of tropical peatlands. Restoration and Wise Use of Tropical Peatland: problems of biodiversity, fire, poverty and water management, International Symposium on

Tropical Peatland, CIMTROP, University of Palangka Raya, Palangka Raya, Indonesia, 73-79.

Wösten, J. H. M., Ritzema, H. P., Chong, T. K. F., and Liong, T. Y., 2003. Potentials for Peatland Development. Integrated Peatland Management for sustainable development - A Compilation of Seminar Papers. , Damai, Sarawak, Malaysia, 233-242.

Wösten, J. H. M., Ritzema, H. P., and Rieley, J., 2008. Requirements for and operational aspects of water management in tropical peatlands. International Symposium and Workshop on Tropical Peatlands 'Peatland development: wise use and impact management', Kuching, Sarawak.

Curriculum vitae

Henk Ritzema was born in Baarn, the Netherlands, on 25 May, 1954. After secondary school he studied Civil Engineering at Delft University of Technology, graduating in 1980. For his MSc research he conducted a feasibility study on a water management system for reusing drainage water in a rice polder in Surinam. After graduation, he started his professional career with the Food and Agriculture Organization of the United Nations as an associate expert and worked on the design and implementation of flood protection and drainage works in Fiji and on the design and implementation of irrigation and drainage systems in the Turkana Province of Kenya. In 1984 he joined the Dutch civil service working for the Directorate-General for Development Cooperation, first as researcher and then as project manager for the Pilot Areas and Drainage Technology Project at the Drainage Research Institute in Cairo, Egypt. In 1989 he joined the former International Institute for Land Reclamation and Improvement (ILRI), Wageningen, and continued working in research, training and consultancy on drainage-related water management for food and ecosystems. His primary interests are the organization, management and implementation of research in drainage-related water management, including capacity building through the dissemination of knowledge. He was editor-in-chief of ILRI Publication 16 *Drainage Principles and Applications*, and co-author of ILRI Publication 60 *Subsurface Drainage Practices*. He was ILRI's Coordinator of Research from 1999 to 2002, when ILRI merged with Alterra. He was Research Leader for programmes on the improvement of waterlogged and salt-affected lands in India and Vietnam and the sustainable management of tropical peatlands in Indonesia and Malaysia. He is the coordinator of ILRI regular training programmes and organizes and lectures in regular and tailor-made courses in the Netherlands (Alterra-ILRI, Wageningen University and UNESCO-IHE) and abroad (e.g. in China, Egypt, India, Japan, Malaysia, Pakistan, South Africa and Yemen). Consultancy activities include missions to China, Egypt, India, Indonesia, The Gambia, Malaysia, Pakistan, Thailand and Vietnam. In July 2008 he joined the Irrigation and Water Engineering Group at Wageningen University as part-time Assistant Professor in Field Irrigation.

Front cover: Martin Jansen, Communication Services, Wageningen University and Research Centre and Peter Stroo, UNESCO-IHE (Photo credit: Ymkje Tamminga).

T - #0082 - 071024 - C16 - 254/178/14 - PB - 9780415498579 - Gloss Lamination